CUSTOM SoCs FOR IoT: SIMPLIFIED

The Emergence of
Custom Silicon
for IoT Devices

Daniel Nenni

Mitch Heins

@2017 by Daniel Nenni and Mitch Heins

All rights reserved. No part of this work covered by the copyright herein may be reproduced, transmitted, stored, or used in any form or by any means graphic, electronic, or mechanical, including but not limited to photocopying, recording, scanning, taping, digitizing, web distribution, information networks, or information storage and retrieval systems, except as permitted under Section 107 or 108 of the 1976 US Copyright Act, without the prior written permission of the publisher.

Published by SemiWiki LLC
Danville, CA

Although the authors and publisher have made every effort to ensure the accuracy and completeness of information contained in this book, we assume no responsibility for errors, inaccuracies, omissions, or any inconsistency herein.

First printing: October 2017
Printed in the United States of America

Edited by: Beth Martin
Graphics by: Aijaz Qaisar

Custom SoCs for IoT: Simplified

PREFACE

Two important trends will be discussed in this book: The disruptive nature of the Internet of Things and the use of the ASIC business model by systems companies to get into the chip business.

Disruptive technology has been an integral part of life at least since the electrical telegraph in the 1700s. My Grandfather, whose lifespan touched three centuries, always marveled at the ice machine in his refrigerator's door, since he delivered ice blocks when he was growing up. Even though he experienced the automobile, air travel, and the beginning of military technology (he was a medic in WWI), the ice machine was always his favorite "gadget." My favorite will probably always be the smartphone because it is the Swiss Army Knife of gadgets.

The big difference between IoT and everyone's favorite gadget is that in the next ten years IoT will touch EVERY gadget in our lives. The Industrial Internet of Things is leading the way with smart buildings, factories, and farms and a very clear return on investment (ROI) as a driver. Smart (autonomous) transportation such as cars, trucks, airplanes, ships, and drones are a close second. Consumer IoT featuring smart home devices and wearables are a distant third because the ROI is not always as clear. Moreover, of course, security is a big IoT market segment in itself and will also touch many other IoT devices as well.

According to Masayoshi Son, Chairman and CEO of SoftBank Group Corp, and Chairman of Arm Holdings, more than one trillion IoT devices will be built between 2017 and 2035. Even more interesting, the market for IoT devices and related services could be worth a staggering one trillion dollars per year by 2035. Clearly, systems companies are in the best position to win this market as long as they can make their own IoT chips and that is where the tried and true ASIC business model comes back into play.

The ASIC business model came about in the 1980s and was the catalyst for what we now call the fabless semiconductor ecosystem. You no longer need the massive capital and semiconductor expertise that is required to make a chip unique to your requirements and keep it out of your competitor's hands. This ASIC revolution not only enabled a slew of fabless semiconductor companies that dominated the semiconductor industry, ASICs also enabled a number of systems companies such as Apple, Cisco, Microsoft, and even Google to become what is now known as "fabless systems companies"—systems companies who control their silicon destiny. Even today the ASIC business is a multibillion-dollar market, and the annual growth rate is increasing rapidly as non-traditional chip companies are getting into the semiconductor design business. Additionally, the make-versus-buy chip decision continues to get easier with very cost effective commercial IP and turnkey ASIC providers like Open-Silicon.

If you look at the product offerings of the top three EDA companies (Synopsys, Cadence, and Mentor), the transition from chip to the system level design is clear. As a result, the sources of EDA revenues are now dominated by non-traditional chip companies that are following Apple's lead. The pure-play foundry business is also experiencing a surge with Apple providing close to 20% of TSMC's annual revenue in 2017.

In the second half of the book, we discuss exactly what it takes to go from specification to chip using the Open-Silicon Spec2Chip Turnkey Solution. Open-Silicon has worked with more than 150 semiconductor and systems companies (large and small) completing more than 300 designs and shipping more than 130 million ASICs, which is why we chose them to partner with us for this book, absolutely.

Daniel Nenni
CEO, Semiwiki
October 2017

CONTENTS

PREFACE i
FOREWORD 5
PART I – IOT DEFINITION 8
   INTRODUCTION 8
   IOT MARKETS 9
   IOT TOP-LEVEL ARCHITECTURE 12
      Edge Devices 13
      Gateway Devices 13
      Cloud Devices 14
      Autonomous Systems 14
      Systems-of-Systems 15
IOT HARDWARE CATEGORIES 16
      Sensing/Actuating 16
      Computing/Storage 17
      Communications 17
   IOT HARDWARE ARCHITECTURES 19
      Edge Device Architectures 19
      Gateway Architectures 21
      Security 22
      Closed Systems 24
      Open Systems and the Role of Evolving Standards 24
   IOT SYSTEM POWER 25
   IMPLEMENTATION METHODOLOGIES 26
      Software 26
      Hardware 29
   ECONOMIC TRADE-OFFS OF DIFFERENT IMPLEMENTATION METHODOLOGIES 30
   TIME-TO-MARKET AND RISK MITIGATION FOR IOT DESIGNS 34
      Platform-based Design 34
      Prototype–to-Production 34
      Turnkey ASIC Companies 35
PART 2 - SPECIFICATION TO CHIP/BOOT 37
   OVERVIEW OF SYSTEM DESIGN PROCESS 37
      Product Specification 37
      System Architecture & Partitioning 38
      SoC Design and Implementation 54

Tape out Data Preparation 65
Mask Data Preparation 65
Wafer Manufacturing 66
Assembly and Test 67
Board Design 67
Silicon Bring-up 72
Production Board Assembly and Test 73
SPEC2CHIP CASE STUDY 74
Voledia Architecture & IP Qualification 74
RTL Design, Integration & Verification 77
FPGA Prototyping & Software Development 77
Physical Design, Manufacturing, Packaging and Test 77
HW Boards & Silicon Validation 78
SMART CITY IOT GATEWAY PLATFORM BASED ON VOLEDIA SOC 80
EPILOGUE 82
ABOUT THE AUTHORS 84

FOREWORD

Enablers of the Internet of Things (IoT) are improving the growth rate of the semiconductor industry in a significant way. Technology advancements in algorithms and processing units have made human-to-machine communication a reality. We are now entering an era where incorporating this capability in smart devices has the potential to simplify, enhance and even save lives. The IoT ecosystem is a symbiotic collaboration of hardware and software developers, building block (aka IP) providers, architects and visionaries who want to translate complex human functions (such as voice, vision, and thought) into simpler, machine-decipherable functions. At the core of this effort are the custom system-on-chip (SoC) solutions that enable designers across vertical markets to meet the performance, power, price, and time-to-market constraints of the quickly evolving IoT universe.

The semiconductor ecosystem has categorized the IoT space into three distinct segments: IoT cloud, IoT gateway, and IoT edge. This segmentation allows key players to devise strategies and offerings in areas of their expertise, which benefits customers with much-needed competition in each segment. Similar segmentation in the computation world helped create the "WinTel" (Microsoft + Intel) ecosystem, which ruled humanity for decades. Segmentation also helps address new and evolving standards, markets and customers in a rapid response manner. Custom silicon solutions have been deployed on the cloud side of the IoT for many years, specifically in networking, telecommunications, storage, and computing. However, until very recently, custom solutions were out of reach for IoT edge and IoT gateway segments due to cost or lengthy development schedules.

The IoT SoC platform approach has opened up many new use-cases for edge applications. Among them are sensor hubs for industrial applications, including outdoor, factory floor and in-room environmental control. IoT gateway applications are also experiencing rapid growth from the custom IoT SoC platform approach. For example, a well-designed IoT gateway SoC platform can address multiple smart city applications, such as waste management, transport, traffic, parking, lighting, and metering.

Custom SoCs for IoT: Simplified

The custom IoT SoC platform approach can speed custom design, reduce risk and cost, and enable the critical differentiation that customers demand. Quality platform development requires extensive experience and knowledge. Platform creators must think like a system company as well as a startup. They need to consider end-use-cases in the vertical IoT markets while designing an easy-to-use platform. Such developers need to be responsible for the core block and its verification, which allows for the highly customized software drivers to be written and used as the core library.

The use of platforms not only opens the door to faster validation of new designs with very little risk but also allows the visionaries and architects to focus on their end goal, which is to bring product differentiation, more use-cases, more functionality and more ingenuity to the world of IoT.

"Custom SoCs for IoT: Simplified" is the first comprehensive book to explicitly define and detail the various IoT architectures. It covers the multitude of security factors, the power budgets associated with different IoT applications, and many more technical considerations that dictate the success of a custom IoT SoC platform, including but not limited to implementation methodologies, as well as hardware and software tradeoffs. This book also provides a detailed case study of a highly successful approach to custom SoC design for an IoT gateway SoC using Spec2Chip turnkey solutions.

It is important to mention that the implications of the Spec2Chip offerings outlined here extend far beyond IoT cloud, IoT edge, and IoT gateway devices. OEMs in other emerging technologies, such as deep learning, artificial intelligence, virtual reality, gaming and autonomous driving cars are benefitting from this Spec2Chip platform approach. Customers in these markets are collaborating with turnkey ASIC providers so they can scale back on, or even eliminate, the risks and loopholes of a lengthy chip design flow, and focus specifically on the core hardware differentiation IP and end application software that they bring to their innovation.

This book deliberately includes a great deal of data and references to real products. We want you to fully understand and appreciate the scope of the IoT ecosystem and the Spec2Chip platform approach that is fueling its expansion. The goal is for you to take this experience and knowledge and apply it to your personal or organizational design flow. Our sincere hope is that your ideas, combined with the proven design methodologies outlined here, will result in a technological advancement that contributes to the IoT universe and those who live within it.

Taher Madraswala
President and CEO of Open- Silicon
October 2017

PART I – IOT DEFINITION

INTRODUCTION

The Internet of Things, more commonly known as the IoT, has arguably been around since the advent of the internet. In its most general definition, the IoT refers to an ever-expanding network of 'things' that are addressable via the internet. Each thing has its own IP address and can communicate with other internet-enabled devices and systems. Historically the things were primarily comprised of various forms of computers and computing peripherals such as scanners and printers. With the advent of cheaper and more capable integrated circuits for sensing and communications and the explosion of wireless communications and mobile devices, the IoT took on its own identity. It has grown into a wide range of devices that can collect information about their environment and communicate that data along to other devices and into the cloud for processing, analysis, and actionable follow-up.

Depending on whose reports you read, the number of connected devices, 'IoT things', is projected to grow to almost fifty billion devices by 2020. In fact, we already have more connected things than we have people on the planet and those connected things will literally be found in every facet of our lives. Per a 2015 report from Harvard Business Review (http://bit.ly/1LXG2NM), there will not be a single industry that won't benefit from the IoT. The IoT space is projected to account for trillions of dollars of GDP within the next ten years and is set to profoundly impact the core strategies and business models for companies around the world. It will make us rethink how we view and gather insights on markets, how we develop products, and most importantly how we interact with customers.

Custom SoCs for IoT: Simplified

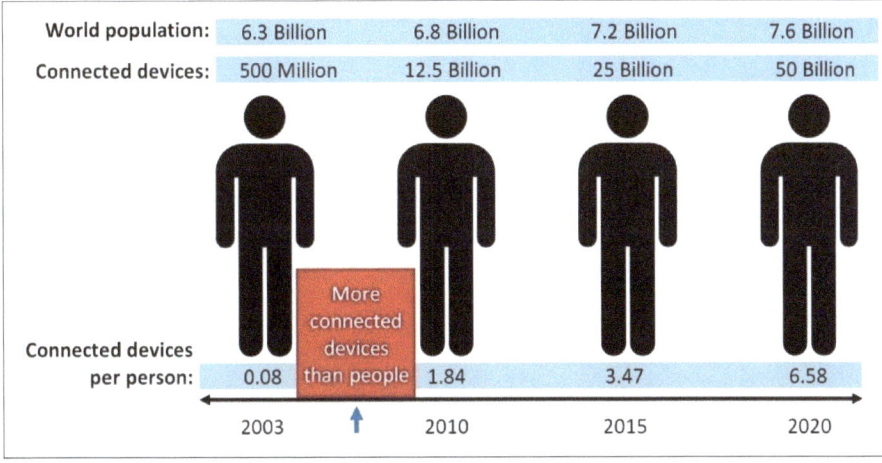

Image from Qorvo, IoT International Conference, Brussels, 2017

IOT MARKETS

The breadth of IoT markets is just beginning to make itself known. Numerous books and articles expound on the idea of smart IoT entities. These entities make use of an array of sensors and actuators, known as edge devices, which are connected to smart gateways and then on to the cloud where any number of applications await. Smart entities include ideas such as smart cities, smart power grids, smart homes and appliances, smart lighting, smart transportation and traffic systems, smart buildings, smart factories, smart automobiles, smart health-care, smart consumer and retail stores, smart agriculture, and just about anything else smart you can imagine.

In the beginning, the idea was as simple as using the internet to do asset management and tracking. Things were tagged with passive RF-IDs and then tracked using an array of sensors to manage their whereabouts and numbers for inventory management. The appeal was obvious for warehouses, clothing, retail, and even grocery stores where large numbers of things were continually being shipped in and out, and inventory management was crucial to the success of the business. The technology quickly leaked over into all sorts of industries where parts and inventory management are a daily part of life.

Asset tracking and smart logistics. Image from WIREPAS, IoT International Conference, Brussels, 2017.

The next big IoT wave came with the ability for things to actively communicate via wireless connections to the internet. The entire paradigm shifted from things that could be sensed to things that can send their own information. Not only could things communicate their location or existence, but they could also communicate information about their status, as well as the status of the environment in which they found themselves. Additionally, this new communication link also provided for two-way communication. The edge devices were uniquely addressable, and messages could be sent directly to them via the internet. Not only could sensors be used to gather data, but actuators could be employed as edge devices to take remote action.

Machine-to-machine communications first started to enable factory lines to be able to self-monitor and more efficiently move items through the line. However, after the advent of the cloud and the ability to address each thing, remote action could also be taken at the behest of a user through a cloud-based application that could communicate to devices. Suddenly we found that we could control the lights in our house while we were on vacation halfway around the world, or we could start and warm up our car on a cold winter's night while sitting in an airliner that was still taxiing its way up to the airport gate.

Custom SoCs for IoT: Simplified

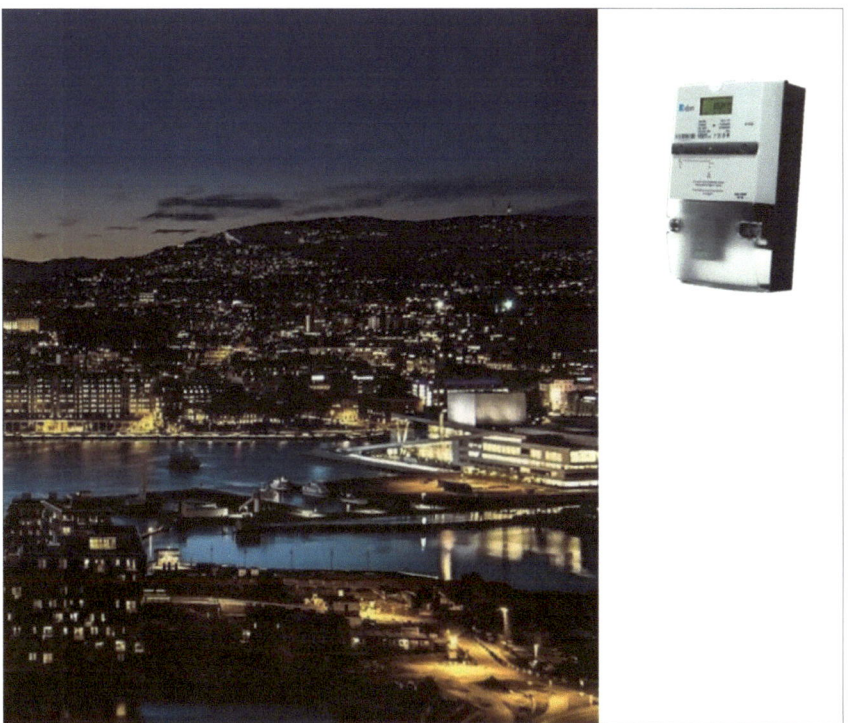

700,000 e-meters in one deployment in Oslo, Norway. Image from WIREPAS, IoT International Conference, Brussels, 2017.

The most exciting IoT advancement now is the advent of truly smart collections of IoT edge devices. These are devices that have enough compute power to enable them to make decisions as a group and to then autonomously take actions. Similar solutions are being investigated for a wide variety of markets as previously outlined.

As an example, grocery stores are already enabling customers to shop online and then not long after, drive up to the store and have the grocery employees bring out their shopping requests right to the car with the transaction having already been paid for through a web application. One can imagine doing the same, but instead of the customer going to the store, the store simply delivers the groceries to your home using an autonomous self-driving car. Appliance manufacturers envision those same groceries being automatically scanned and inventoried as they are put away into IoT-enabled refrigerators, and pantry stores complete with registration of expiration dates and recall codes in case a bad batch of beans slips through an IoT-based food inspection line. As inventories in the refrigerator or pantry run low, a deep neural network analyzes what needs to be re-ordered and starts the whole process over again, with or without human intervention.

One can envision entire ecosystems linked via IoT systems-of-systems with feedback loops and autonomous communications between systems initiating actions and responses. This paradigm is set to play out repeatedly in a variety of marketplaces such as medical, industrial, transportation, energy, agriculture and big data analytics.

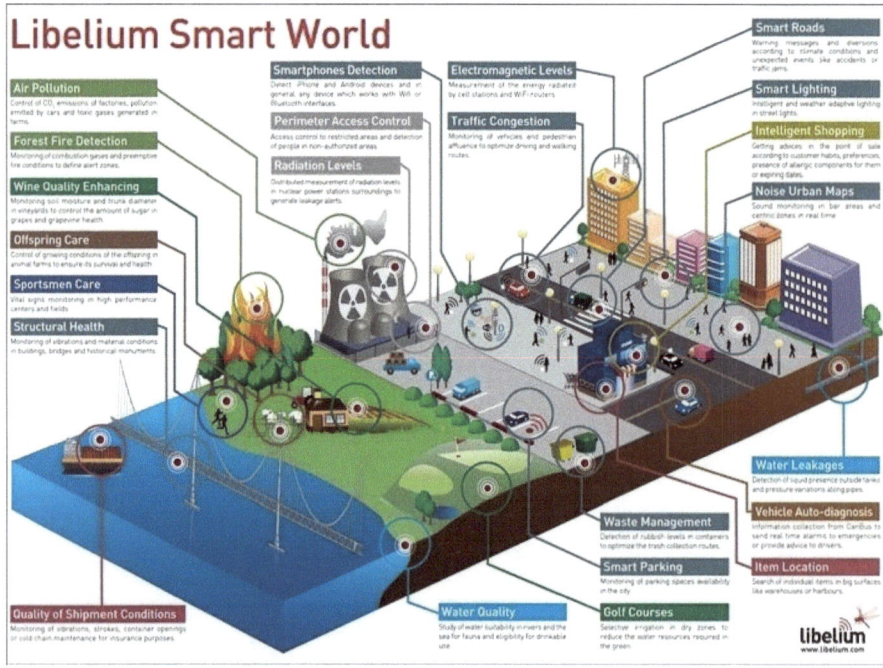

Libelium Smart World. Image courtesy of Libelium.

IoT Top-Level Architecture

As can be inferred from the previous scenarios, there are many complexities to be managed. At a high level, IoT solutions tend to be comprised of three main architectural stages. These stages start at the edge of the internet and include what are known as 'edge' devices. The edge devices next talk to the second stage known as IoT gateway devices. Finally, the gateway devices then talk to the third stage which is known as the cloud.

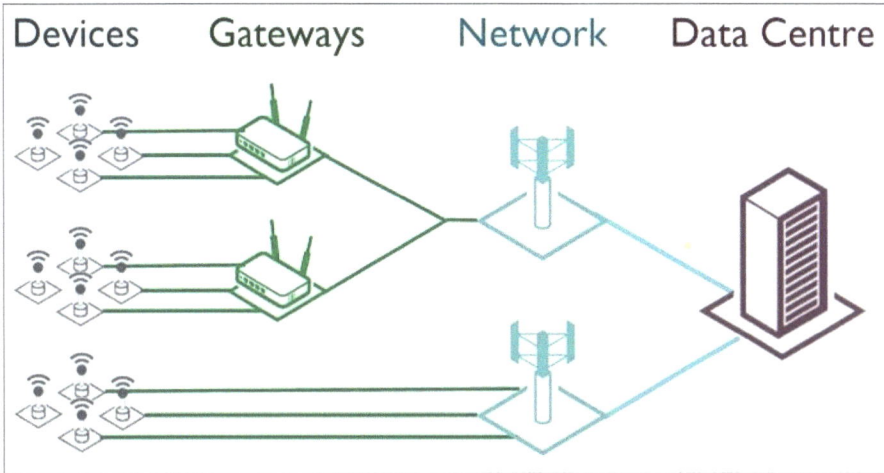

IoT system stages. Image courtesy of ARM.

EDGE DEVICES

Depending on the IoT application and vertical market, edge devices may be relatively simple with only one or a few sensors. A good example of this would be a patient who is wearing a bio-sensor that is sending raw telemetry data to the cloud for inspection by a doctor's office.

Alternatively, the edge devices can be very complex, integrating many different sensors (known as a sensor hub) and local processing units being used for data fusion and data analysis. An example is a traffic control system or an autonomous vehicle. In both cases, the edge devices will have some type of communication hardware that enables them to talk to the internet, usually through a direct wired connection or one or more variants of a wireless link.

GATEWAY DEVICES

IoT gateways connect the edge devices to the cloud. These devices are routers in the traditional sense in that they are controlling communications with the IoT edge devices. The gateways, however, are more than just routers. Their job is to establish and maintain secure, robust and fault-tolerant connections with the edge devices.

Communications with edge devices is a two-way phenomenon. Information flows up from the edge devices and can be either aggregated and analyzed at the gateway or be pushed onto the appropriate cloud-based storage for later data analysis. Conversely, data can flow down to the edge devices from the gateways. This data may be as simple as a basic device health-check or software updates to the edge device. Alternatively, the data could be complex sequences of commands used to communicate with actuators on the edge devices.

IoT gateway devices use to connect edge devices to the internet. Image courtesy of ARM.

CLOUD DEVICES

While IoT gateway devices oversee multiple edge devices, from there, an IoT system may employ network appliances that oversee many IoT gateways and manage data traffic to and from servers that will be used for data analytics and visualization. Depending on the type of IoT system architecture employed (centralized or decentralized), there may be dedicated network appliances and servers tied to the system. Alternatively, in a decentralized system, non-dedicated appliances and servers may be used with a layer of software above them that helps to coordinate the data storage, analytics, and visualization.

AUTONOMOUS SYSTEMS

A derivative of the classical IoT system that sends data to the cloud for processing is a more autonomous system with IoT gateways that act as decision-making devices that aggregate data from many different edge devices and then broadcast commands back to those devices with actions to be taken. The traffic control system might be a good example of this. The idea here is for an interaction between drivers, smart vehicles, and city resources such as availability of parking spots, collection of parking fees, logistics and fleet management for delivery companies, traffic congestion and mitigation monitors, electronic toll collections, roadside assistance, and autonomous taxi services.

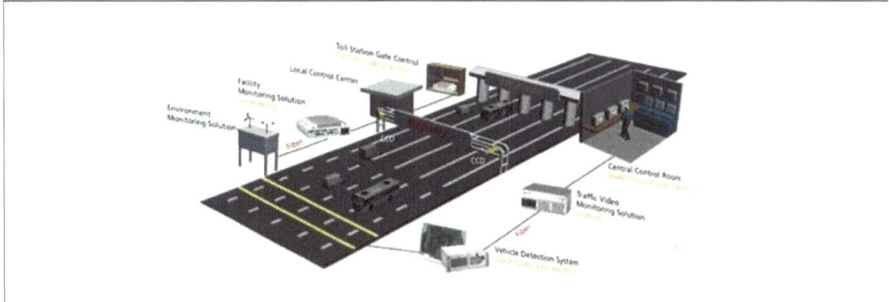

Example of smart city IoT-based traffic services. Image courtesy of Neetikasyal.

SYSTEMS-OF-SYSTEMS

In some cases, sensors, and actuators talk to more complicated compute engines running on IoT gateways. In other cases, sensors, actuators, and dedicated hardware processing engines are being integrated into single, custom application specific integrated circuits (ASICs) known as systems-on-a-chip (SoCs). Embedded software running on these ASIC SoCs are using sophisticated artificial intelligence (AI) algorithms, known as deep neural networks (DNNs), to fuse and analyze environmental sensor data. The results of the analysis are used by the system to make complex decisions and initiate actions through remote actuator edge devices.

Furthermore, complex systems-of-systems are being built from these highly integrated collections of IoT subsystems. These larger systems-of-systems are still in the conceptual stage, but we are seeing the beginnings of them in the likes of driverless cars where several different and varied sensors, microprocessor compute engines and embedded software are used along with actuators for braking, steering, and engine control to drive the car. It is envisioned that the autonomous car itself will one day become the equivalent of a complex IoT edge device. It will communicate with other edge devices and gateways, such as other smart cars and smart city traffic control systems, to enable it to navigate through the city while helping the city maintain optimal traffic patterns and avoid traffic jams.

IoT Hardware Categories

Hardware used in an IoT solution can be broken into three categories:

- Sensing/actuating
- Computing/storage
- Communications

Sensing/Actuating

Sensing and actuation are very application specific and can fall into a wide range of capabilities and categories. They typically fall into machine equivalents of our five senses of sight, sound, smell, touch, and taste. Machine vision and image recognition, acoustic monitoring and filtering, biometric and biological sensing and tactile information such as temperature, pressure, acceleration, location, proximity, and orientation are all examples of the types of sensing that are being pursued. These sensors go well beyond the human capability to measure with our five senses in terms of breadth, accuracy, speed, and longevity and we are just now scratching the surface of what can be done.

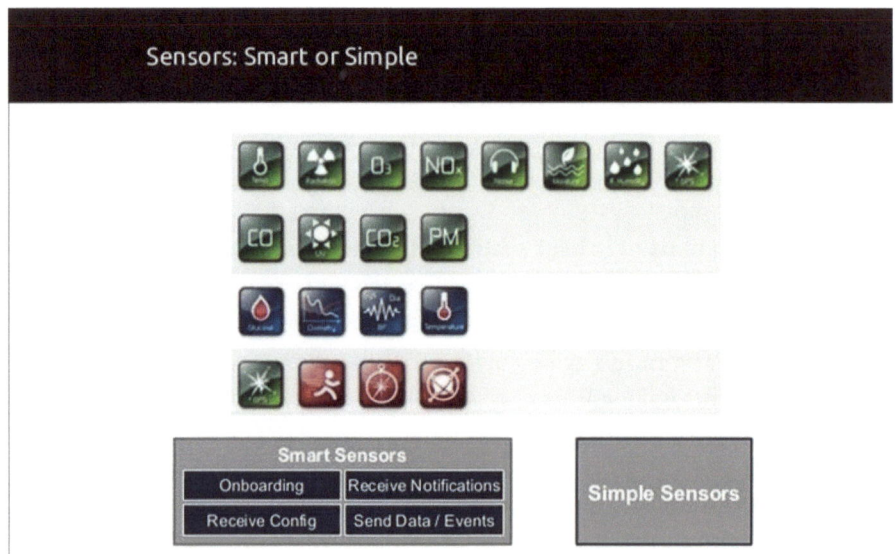

Sensors, smart or simple. Image courtesy of Mulesoft, IoT Expo, 2014.

COMPUTING/STORAGE

In addition to sensing, IoT systems can also be used to process and store data both for immediate responses to stimulus and for longer-term analytics that can be used to study data over an extended period and breadth of conditions. For almost every type of sensing condition there now exists specialized computing engines, or cores, that process the raw data to make sense of the information coming in from the sensors. These cores include microprocessors, microcontrollers, digital signal processors, graphics processors and image processors to name a few.

COMMUNICATIONS

Like sensing, the type of communications standards employed are highly application dependent. Requirements for how one communicates with the edge devices and gateways will be shaped by the use model and architecture of the IoT system. Besides power, specific requirements for latency and response times will have a big impact on how much or how little local analysis is done in either the edge devices or the gateways. IoT systems that need low latency, rapid responses will likely have more compute power and local memory than those systems that might feed a business intelligence/data analytics system where sensor data may be fused and analyzed in a distributed fashion within the cloud. The applications drive how far and how much data must be communicated as well as determining speed and latency requirements for the communications.

A presentation by Qorvo at the 2017 IoT International Convention in Brussels, Belgium gave a good example of the breadth of communications standards being used. Which standard is employed depends on the distance data must be communicated versus different applications, some of which were sensitive to high data volume rates while others are more focused on low power/long battery life. Distances for communication were broken into three categories: peripheral connectivity (less than 10 meters), local networking (10-to-100 meters) and wide area networking (distances in kilometers).

Custom SoCs for IoT: Simplified

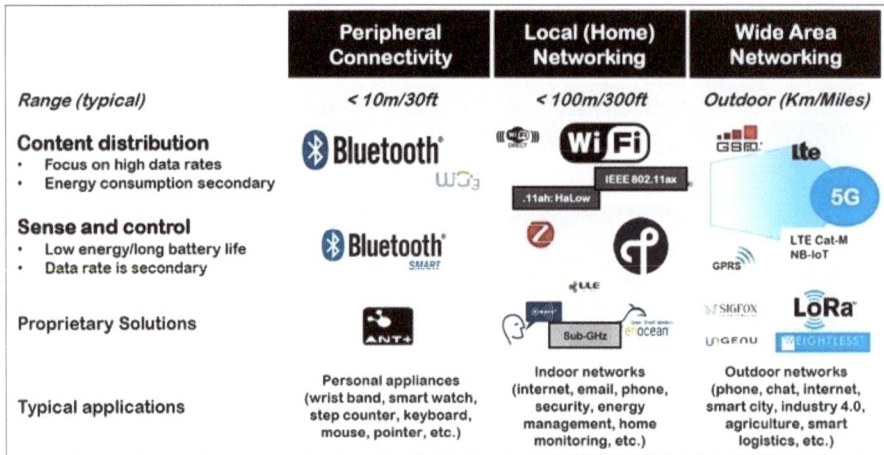

Connectivity standards for IoT designs. Image courtesy of Qorvo, IoT International Conference, Brussels, 2017.

For IoT systems needing higher data rates, some of these standards may apply:

1. Peripheral connectivity (< 10 meters)
   Bluetooth and WiGig. WiGig uses the 60GHz band as compared to regular Wi-Fi, which operates at 2.4 GHz and 5 GHz. WiGig is meant to enable multi-gigabit performance for data-intensive applications over short distances.
2. Local networking around the home (10-to-100 meters)
   Wi-Fi or Wi-Fi Direct and WirelessHART. Wi-Fi Direct allows a direct Wi-Fi connection between devices without the need for a wireless router. Also available are 802.11ah (HaLow), which is an alternative for faster connections but with less power consumption than the newer 802.11ax for higher bandwidths.
3. Wide area networking (kilometers)
   This category has many possibilities, including GPRS, GSM, LTE, LTE Cat-M, narrowband IoT (NB-IoT), and 5G. 5G stands to be a prominent choice over time as it becomes ubiquitous for cellular users. It may also be the workhorse for MMTC (massive machine type connections).

For battery powered IoT systems that are looking to conserve energy, and where the data rates are secondary, these standards may apply:

1. Peripheral connectivity (<10M)
   Bluetooth Smart or Bluetooth Low Energy (BLE)
2. Local networking around the home (10-to-100M)
   Zigbee and Zigbee Mesh, Green PHY, and possibly HaLow.
3. Wide area networking (kilometers)
   GPRS, LTE Cat-M and NB-IoT, and again the possibility of 5G just because of the infrastructure build-out that will likely take place.

In addition to the competing standards, there is also a host of proprietary communications solutions such as ANT+ for peripheral distances, Sub-GHz, EnOcean, Z-Wave, and solutions from the ULE Alliance for local networking and SIGFOX, LoRa, INGENU and Weightless for wide area networking.

IoT Hardware Architectures

One thing common to all IoT-based systems is that a full system architecture must be thoroughly planned, both at the hardware and software levels, to ensure that the correct types of edge and gateway devices are designed. There are many design trade-offs to be made when it comes to how much data should be transferred at all levels of the data flow between edge devices, gateways, and the cloud. These decisions will have a profound impact on specifications for hardware device configurations in terms of compute capabilities, local memory, and communication capabilities. These decisions will drive requirements for device power budgets, which will in turn drive decisions about how the devices will be powered.

Architectural decisions tend to be very application specific. Imagine two sensing devices, one found on a factory floor and the other in the form of an ingestible pill. The former may be powered directly from a wall plug while the latter may get its energy from a kinetic energy harvesting MEMS circuit.

Edge Device Architectures

IoT solution providers have a wide range of hardware capabilities from which to choose to accomplish the functions of these three categories. Initial solutions started with standard product parts individually purchased and assembled into systems using printed circuit board technologies. As solutions become more prevalent and competition becomes more fierce, solutions providers are pushing to integrate one or more of these functions onto an IC to enable further cost reductions, additional functionality, and better reliability.

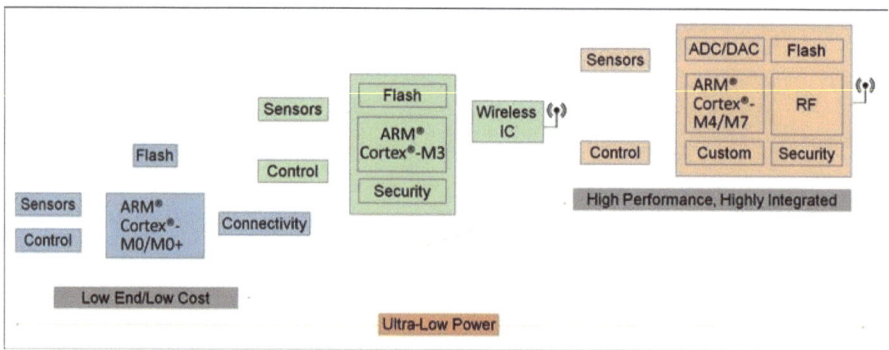

IoT edge device architecture. Image courtesy of Open-Silicon

DATA COLLECTION

At its most basic form, IoT edge devices may simply be doing data collection using one or more sensors connected to the internet either through wired, or now more predominantly, wireless communications. Initially, the architecture for these devices was simple, with only the sensor and a controller with a communications interface to enable data to be sent directly to a gateway in real-time. A good example is the early surveillance systems that let you know if a door or a window had been opened. In these systems, two-way communications were very limited, usually consisting of a simple handshake between the sensor and the hub to register that the sensor was online and reporting to the right hub.

With the advent of security concerns, these devices are slowly being replaced by smarter versions that can handle multiple different sensors and actuators while also having dedicated hardware to secure the devices from being used in denial of service attacks.

DATA COLLECTION AND SIMPLE ACTUALIZATION

The next step in complexity was the addition of more complex two-way communications to enable both data collection and actuators that could be controlled by a remote source. Early examples would be the use of surveillance cameras that could be panned and zoomed by the remote user through a phone app. The app used the cellular system to talk to the IoT system app running in the cloud. The system app then sent instructions through a wireless gateway hub to the edge devices. In these cases, the edge devices would have a controller, sensor interfaces, communications hardware and possibly some local memory that would be used to store time-spans worth of data that could then be retrieved by the user.

DATA COLLECTION, DATA FUSION, AND DATA ANALYTICS

As we progress up to higher levels of complexity, the edge devices start to take on more data analysis and data fusion functions, breaking down data into useful information before it is sent along to the IoT gateway. In these smart devices, we see sophisticated system-on-chip ASICs—possibly with multiple microprocessor cores, each tuned to a specific data analysis task— along with local on-chip memory, communications interfaces for different sensors, dedicated hardware for root-of-trust for validation, and encryption hardware for secure communications.

GATEWAY ARCHITECTURES

IoT gateways provide connectivity for edge devices to send and receive information to/from the cloud and other edge devices. They are also used to manage the devices, including bidirectional routing of data and control messages, provisioning of edge devices to specific tasks and in some cases, data fusion and analytics for collections of edge devices for which they are responsible.

Because these devices are truly the gateway between the edge devices and the internet, security is a must and, as such, these gateways will have dedicated hardware to manage secure computing and communications. Depending on the number and types of edge devices for which the gateway is tasked to manage, the gateways will also have dedicated hardware and software to manage communications with these devices using different standards and protocols.

In addition to acting as a hub for multiple sensors, IoT gateways may also act as a hub of hubs. Each of the children hubs may be responsible for handling edge devices of a specific kind. Alternatively, the division of hubs may also be by the expected distance to be crossed between the hubs and sensors, with a different hub being customized to handle different communications standards for different distances as discussed earlier.

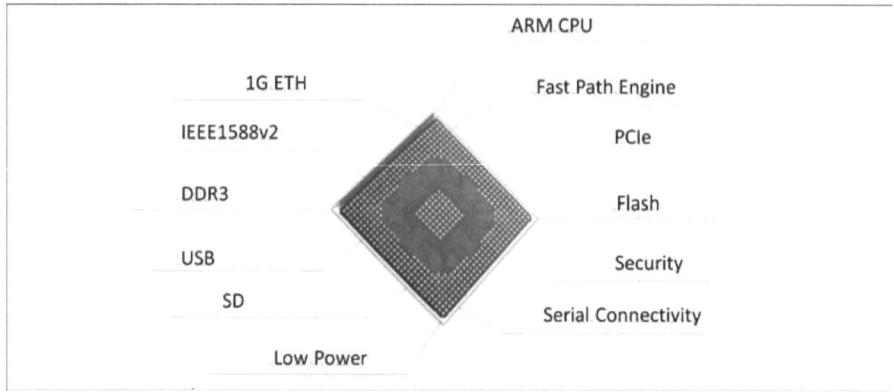

IoT gateway features. Image courtesy of Open-Silicon.

SECURITY

IoT security features can include the following:

- Root of trust and secure boot
- Isolated execution environments
- Encryption
- Tamper protection

ROOT OF TRUST AND SECURE BOOT

One of the most important factors for ensuring an IoT secure system is what is known as RoT or Root of Trust. In IoT systems, there are many ways to establish RoT. For edge devices and IoT gateways, the most secure way is to use some dedicated hardware that has been hardened so that it is highly unlikely to be compromised, or it is set up so that it can't be modified at all, or it can't be modified without cryptographic credentials.

ISOLATED EXECUTION ENVIRONMENTS

A second factor for ensuring IoT security is to close holes in the attack surface that might allow a nefarious actor to insert itself into the edge or gateway device and gain control. The best way to ensure this doesn't happen is to secure key device functions and controls by isolating the execution environment that executes these functions. The latest versions of the ARM V8 M33 and M23 cores are excellent examples of using isolated and separate execution environments to handle secure device bootup sequences, encryption, and over-the-air (OTA) firmware updates. In these cases, provision has been made in the cores to totally isolate key processor registers and memory from the outside world unless instructions come through encrypted channels that have been verified using the RoT.

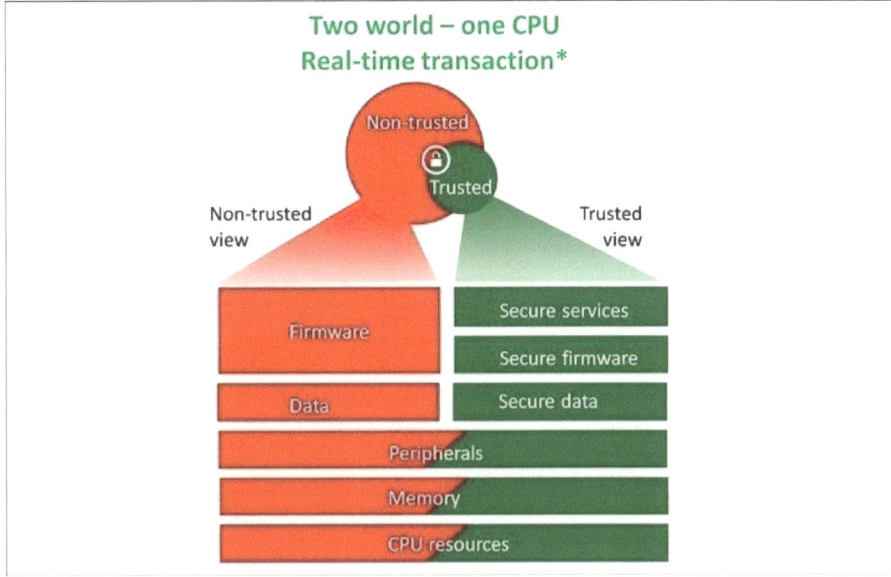

ARM TrustZone for ARMv8-M. Image courtesy of ARM.

ENCRYPTION

IoT devices must be enabled to be secure by default. Since these devices must communicate with the outside world to fulfill their function, they must have a way to both establish secure links and to protect data that could be intercepted in transit between the devices, gateways, and the cloud. Most solutions are a combination of both hardware and software. Hardware is used to generate crypto keys and software is used to decide which data is to be encrypted when being sent.

Standard encryption algorithms—such as DES, TDES, and AES (with 128, 192 or 256 bits)— are now commonly used along with hardware approaches that include message summaries with the communication that can be checked to make sure the data has not been tampered with along the way.

Algorithms for this type of message authentication include MD-5, HMAC, SHA-1 and SHA-256, and AES-GCM.

TAMPER PROTECTION

More recently, hardware vendors are now including specialized hardware in their ICs that can detect attempts to tamper with the device. This might include out-of-range sensors for voltage, temperature, clocks, and the use of test modes without an authorized verified command enabling them. Depending on the IC, if tampering is detected the device may be taken offline or disabled and key memory areas may be reset to thwart attempts to copy or change the secure data.

CLOSED SYSTEMS

We are in the very early days for IoT and markets are still evolving with many opportunities. Complicating matters is that there is no one IoT market, but instead, there is a heterogeneous landscape of markets with a wide variety of both business and technical requirements. Initially, these markets will be addressed with closed proprietary systems that address very specific value propositions in their vertical market segments. Successful solutions will grow and providers of these solutions will naturally look at adjacencies within their markets to exploit.

At some point, the closed systems will eventually bump into other proprietary systems and it will be necessary for those systems to communicate. Initial market pressure will drive competitors to collaborate loosely, and de facto standards for interfaces between systems will form that will create the basis for eventual open standards.

OPEN SYSTEMS AND THE ROLE OF EVOLVING STANDARDS

Over time, the marketplace will sort out the winners and the losers and solutions providers will start to look for the next level of integration to expand functionality while simultaneously reducing costs. Interface standards will emerge at multiple levels that will enable IoT systems to communicate and collaborate with each other. Technologies are already competing for these coveted standards at the physical link layer and at the network transport layer. Eventually, standards will also appear at the application layer, although at this level they will tend to be market application specific.

Initially, IoT solutions may look to differentiate based on various technical aspects of how well they can serve their markets. However, as standards appear, these technical differences tend to get equalized out of the value proposition, and the IoT solutions provider must look to a higher level of differentiation. At first, standards would seem to stifle innovation as they tend to commoditize solutions players at the level of standardization. In the end, this also works to stabilize offerings at the level of standardization and frees up solutions provider resources to look for the next level of differentiation, which starts another cycle of innovation and competition.

The key for IoT solution providers is to make sure they have a flexible, configurable solution that can pivot with the market as the standards solidify and become real. It's a very costly mistake to architect a solution that bets on only one of many variations of a would-be standard. If the wrong variation is picked, or worse yet the wrong standard, a solution could be dead in the water before it even hits the market.

IoT System Power

Much has been written about the need for lower power IoT devices. While it is generally true that low power consumption is better, it should be noted that this is an area where one IoT size does not fit all applications. As an example, wearable IoT edge devices have a need for low power as they will likely run on batteries that are expected to be recharged no more than once every few days. An exercise monitor worn on the wrist is a good example of this. With clever design, the application may even be able to do a bit of energy harvesting from the motions it is built to monitor to enable longer battery life. These applications do not necessarily need to be in transmit mode except for when being polled by a web application to upload their information. Shutting off the circuitry to transmit until it is needed can save these devices a lot of power. Depending on what the application is doing, it also may not need a very fast duty cycle to be able to measure its environment.

Alternatively, some edge devices will be plugged into a continuous source of power, so while they should be efficient and not burn too much power, they do not need to be nearly as stingy as their wearable counterparts. Good examples of this may be smart meters for houses or smart sensors in a factory assembly line. Again, depending on what is being measured, the duty cycle will vary greatly.

Gateway devices will also have significantly different power budgets depending on their intended function and the types of edge devices for which they are responsible. Some gateways may be dealing with very large amounts of data and will require significant speed to be able to process and fuse data from multiple different sensors in real time. The amount of data to be sent, along with required latency and quality of service metrics, will significantly impact IC architectures and the power that may be consumed by them.

In general, clever architectures will use both software and hardware to manage to their power budgets. Data gathering devices may only sync up to gateways at certain intervals, so much of the IC logic that handles the data manipulation, transmission, and reception can be put to sleep during the periods when the device is only gathering data.

The amount of data transmitted and the distance it needs to be transferred will greatly effect consumed power. Long distance transmissions are expensive from a power viewpoint, so the IoT system designer must carefully consider trade-offs while designing the overall architecture of their systems. It may be cheaper to have a local gateway fuse and crunch data and send only metadata up to the cloud as opposed to sending gigabytes of raw data. The type of hardware used to do this crunching can also make a big difference in the power consumed. When doing design trade-offs, the system designers need to match up the types of work to be done to the types of processors that can efficiently do the job while considering where data fusion and analytics are to be done.

IMPLEMENTATION METHODOLOGIES

SOFTWARE

As with all systems, there are many different trade-offs that must be made when deciding how best to partition the system between functions done in hardware and those done in software. All IoT systems use a combination of hardware and software, and they use them at multiple different levels from the edge devices to the gateways and into the cloud. Each of these devices will have a software stack that provides, at a minimum, hardware-based security and communications protocols to allow the devices to communicate with each other.

Since the devices are interacting in real time, IoT edge devices will typically have some type of RTOS (runtime operating system) that is used to handle interrupts, task scheduling and network handshaking between other edge devices and the gateways that manage them. The RTOS must also be able to handle OTA (over the air) updates that are sent to them from the gateways while also ensuring secure connections are maintained and messages have not been compromised by outside agents before acting on them. Depending on the complexity of the edge device, the RTOS will also need to provide services that can be used by application programmers to control and modify the behavior of the device.

First-generation IoT systems may have a centralized architecture where IoT gateways and cloud-based applications specifically target an end application where they are in control of a limited number of sensors. Good examples of this are sensors in a factory assembly line. Typical operating systems for these types of designs include ARM mbed, FreeRTOS, Nucleus, ThreadX, uCOS, and Unisom, depending on the processors being used in the gateway ICs.

Later generation IoT systems are starting to adopt the idea of distributed sensors networks. These networks can be a combination of 1000s of sensors that work together in a self-combining semi-autonomous way to sense, process and then act on the data. Wireless sensing networks also take on an additional function of having to optimize resource utilization based upon real-time changes in the environment. Examples of real-time wireless sensing network operating systems include Contiki, LiteOS, Mantis OS, Nano-RK, RETOS, SOS, and TinyOS.

Similarly, gateways have their own software stack and operating systems that must manage the multiple interactions between the edge devices and the cloud. Depending on the amount of data fusion and data analytics to be done at the gateways, the gateway operating system can be more complex, as data fusion and analysis usually implies multiple heterogeneous cores that will be doing parallel processing of data. In these cases, each of the unique types of cores may be running their own operating systems. This implies architectures that need coherent memory access strategies and sophisticated on-chip communications with a master operating system that uses a network-on-chip to coordinate movement of data between cores and off-chip to the cloud or edge devices.

The communications software stack is furthest along regarding standardization. However, there are several competing standards that will take some time to settle out. At a minimum, most IoT developers agree on the following communications layers. A list of these layers from top (application) to bottom (datalink/physical) is shown along with some of the competing standards being deployed at each layer.

- Application/data layer
  Standards include HTTP, CoAP, MQTT, etc.
- Transport layer
  Standards include TCP and UDP
- Network layer
  Standards include RPL, IPv4, IPv6, 6LoWPAN, all with TLS and DTLS security
- DataLink/physical layer
  Standards include 802.15.4, Bluetooth/Low Energy, RFID/NFC, Wi-Fi, 3GPP, etc.

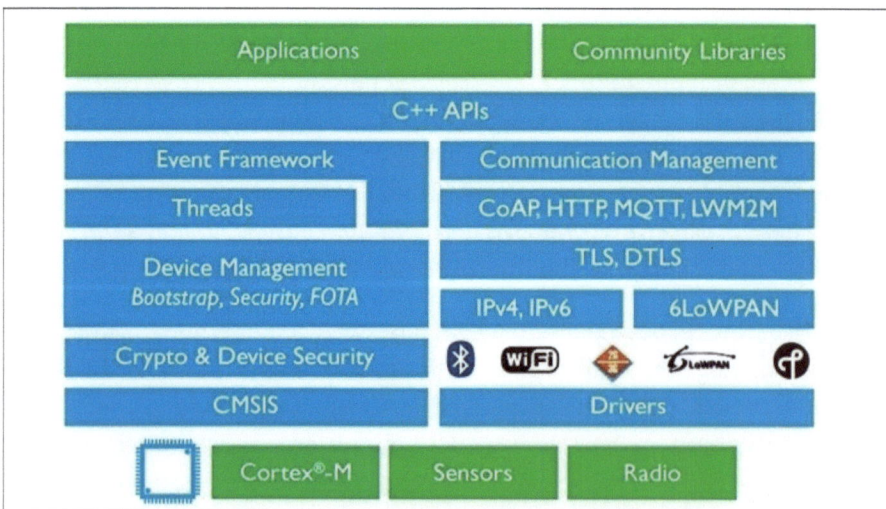

ARM mbed operating system stack for IoT devices. Image courtesy of ARM.

HARDWARE

IoT edge devices and gateways can be designed using a variety of methods including off-the-shelf ICs such as microcontrollers, processors, or field programmable gate arrays (FPGAs) integrated onto a board or into a system-in-a-package (SiP). Different applications have different requirements, and not all features will be available in one single device. Alternatively, designers may choose to create custom system-on-chip (SoC) IoT ICs. In this case, the IP blocks required for the application are selected and stitched together on a single chip.

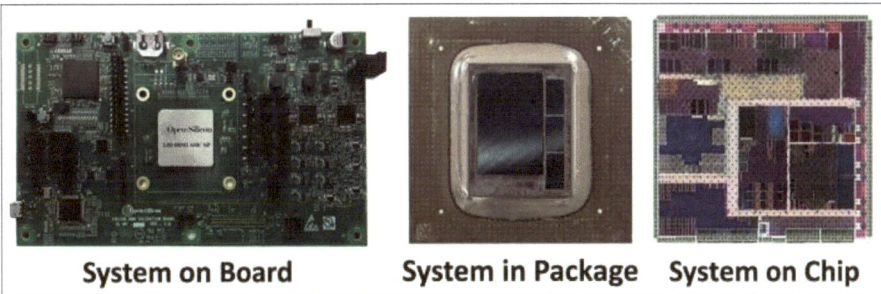

System on Board   System in Package   System on Chip

Hardware options. Image courtesy of Open-Silicon.

To help the hardware/software co-design process, many times these custom ICs will be prototyped by combining FPGAs and standard processor cores with specialized analog components on daughter boards. These environments can be used stand-alone or in conjunction with hardware emulators (large systems that include multiple FPGAs) to work out system bugs. The advantage of these systems is that they are many times married to software development environments that allow software engineers to write code and download it directly onto the memory on the daughter boards for hardware/software debug. The functionality is verified with actual peripheral components used in IoT edge and gateway designs. These prototypes can be used to sort out a variety of features and problems and allow the system to be demonstrated to potential customers before committing designs to production. Once the system is proven to be functioning as planned, the designers can then migrate the FPGA design to a full custom IC.

For many companies, creating an SoC can be a daunting task. However, that need not be the case. There are companies, such as Open-Silicon, who specialize in turning customer ideas into system-optimized custom IC solutions for their customers. Businesses like Open-Silicon are responsible for every stage of the IC design and manufacturing process including system and software design, IC architecture and partitioning, logic design, and physical layout. Open-Silicon goes a step further by partnering with manufacturing companies to manage the manufacturing, packaging, testing, board design, and final bring-up of ICs and boards for their customers. To facilitate this process, Open-Silicon uses a platform-based design methodology that provides a tremendous advantage for IoT systems designers. These pre-configured platforms can comprehend the use of multiple different processing cores and communications protocols. The same platform allows designers to trade-off different processing cores or even allow for integration of custom IP built by the system designers. The platforms will have also considered device and gateway security and provided hooks and mechanisms to ensure that the system can be made secure.

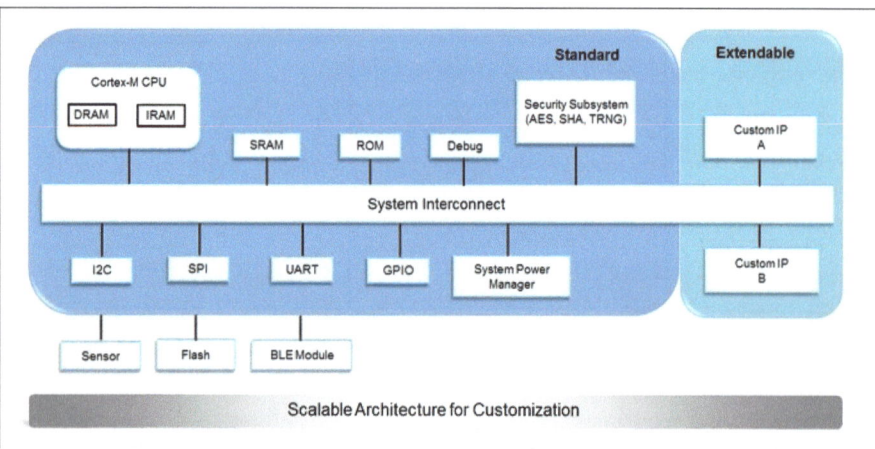

An IoT reference platform with an ARM Cortex-M CPU, memories, and various interface blocks. Image courtesy of Open-Silicon.

ECONOMIC TRADE-OFFS OF DIFFERENT IMPLEMENTATION METHODOLOGIES

Once a good definition of the overall system is in place, designers must decide on how to best partition the design between functions that will be implemented in hardware versus software. From an economics point of view, there are many factors that must be taken into consideration. In general, specialized custom hardware is desired when higher bandwidth and processing speed is required and when security demands hard encryption. Eventually, the final production system will be built using a combination of standalone components and either ASSPs (application specific standard products) or custom IoT SoCs. The decision about which form to use is usually driven by volume expectations and the costs of the BoM (bill of materials), which is the list of various components that make up the entire system. Custom IoT SoCs can also provide for specialized hardware that may be used to differentiate the design and for which standard solutions may not exist.

Designers analyze and break down system cost into multiple cost groups. The first cost group is what is called non-recurring engineering costs, or NRE. The second cost group deals with recurring costs of manufacturing, packaging, testing, and deployment of each system unit.

NRE is composed of up-front development costs of designing and deploying the system. It includes such items as the initial design and verification of the system, software, IC(s), specialized packaging (ICs and system), printed circuit board(s), prototypes, test harnesses, documentation, certifications, and training. Systems that use custom IoT SoCs typically have higher NRE costs as the custom chips must be designed, packaged, and tested whereas ASSPs that have already been designed have had their NREs amortized over multiple customers who use those products. As an example, typical photomask costs for a custom IoT SoC can run from $500,000 to $1 million (a far cry from $5 million to $10 million mask costs of state-of-the-art server processors, but still, it's a lot of money).

Conversely, custom IoT SoCs typically have lower individual unit costs than their ASSP counterparts as the custom ICs have been designed to have just enough functionality to perform their function, whereas ASSPs are typically larger devices that may have more functionality than required by the application.

A good high-end IoT SoC example of this would be a system that uses a standard DSP with a few other standard chips. The NRE for the standard products solution would be $0, but the BoM cost might be $25/unit. A cost-

Custom SoCs for IoT: Simplified

reduced custom IoT SoC replacement for this system may have an NRE of $5 million but a BoM cost of only $4/unit. At 5 million units, the total cost of ownership (TCO) for the standard product solution would be $125 million (5M units x $25/unit), while the TCO for the custom IoT SoC would be $25 million ($5M NRE + 5M units x $4/unit). This represents a TCO savings of $100 million when using a custom IoT SoC solution over the standard product solution.

A good low-end IoT SoC example of this would be a system that uses a standard microcontroller with eFlash with a few other standard chips. The NRE for the standard products solution would be $0, but, the BoM cost might be $10/unit. A cost-reduced custom IoT SoC replacement for this system may have an NRE of $2 million but a BoM cost of only $4/unit. At 5 million units, the total cost of ownership (TCO) for the standard product solution would be $50 million (5M units x $10/unit), while the TCO for the custom IoT SoC would be $22 million ($2M NRE + 5M units x $4/unit). This represents a TCO savings of $28 million when using a custom IoT SoC solution over the standard product solution. Also, because ASSPs are more general in nature, their performance may not be suitable for the IoT application or as good as the custom IoT SoCs, which are architected for the specific application.

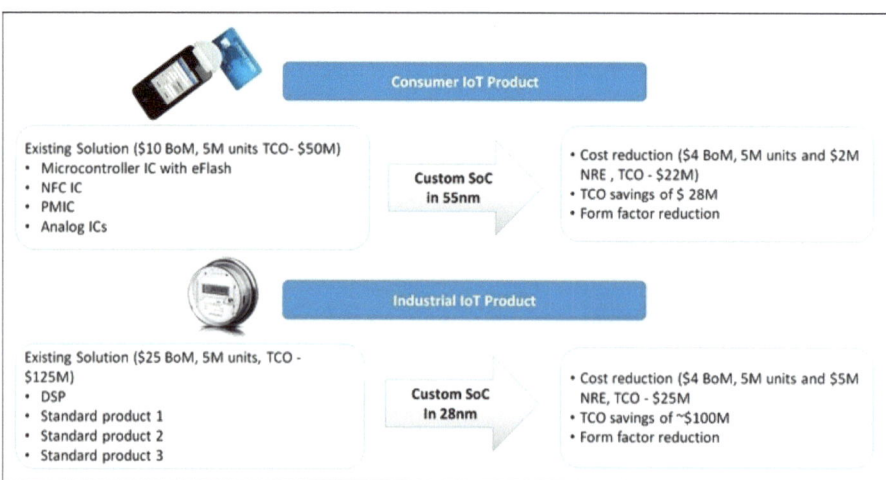

Trade-off analysis for standard product BoM vs. custom SoC solution. Image courtesy of Open-Silicon.

The recurring engineering costs include the incremental costs to manufacture, package, test, and deploy each incremental unit of the system. Recurring costs are highly dependent on the BoM and include the cost of any hardware piece parts that make up the system, including printed circuit boards, individual components, and standard products on the board, and any custom IoT SoCs that may have been designed into the system. The BoM cost also includes costs for assembly and test of each individual deliverable unit of the system. Since these units are expected to ship in the millions of units, attention is paid to trying to reduce or minimize BoM cost while still meeting the overall system requirements needed to meet the application's requirements. It's important to remember that recurring costs may also include license royalty fees for any IP blocks that are used as part of any custom IoT SoC used in the system.

An additional set of costs that must also be considered is the cost to operate and maintain the system over time. As discussed, in fast-changing markets like IoT, it can be expected that IoT systems will need to be agile and able to respond to both a changing physical environment as well as a changing economic environment. Also, because of the number of units involved, the overall cost of power and energy required to run the system must be considered. A power-hungry ASSP-based system may not be economically feasible depending on the application and, as such, architectures that make use of standard products may find themselves being too inefficient for the market.

The last set of costs to be considered involve time-to-market and lost opportunity. These costs are perhaps the hardest to quantify but could be the most significant of the system. Being first to market with a viable product can make a huge difference in market share and, therefore, profit that can be expected from the system. Being first to market with a product that is close and that can be easily configured to match changing market requirements can be a good strategy, but it requires more work in the early system design stages to determine what needs to be configurable and how that configurability will be accomplished. At a minimum, this usually implies some type of capability to provide over-the-air (OTA) software updates for the devices.

TIME-TO-MARKET AND RISK MITIGATION FOR IOT DESIGNS

A key consideration for time-to-market is the time to design, implement and test the system. If the system includes any custom IoT SoCs, one must remember that these devices have longer delivery lead times and can also represent higher design risks for first-time design teams who have never done a custom IC design. There are several strategies to mitigate these risks. Three strategies are covered next.

PLATFORM-BASED DESIGN

The first strategy is to use platform-based design in which the system is built around predefined platforms. Platforms reduce the overall NRE, design risks, and time-to-market, as much of the system has already been designed, implemented, and tested. Platforms are typically built around CPU architectures and system interconnect strategies. A good example of a platform-based design is the Open-Silicon IoT platform. This platform is based on one or more ARM Cortex processors combined with the onboard SRAM and ROM memory, and a variety of possible communications protocols such as I2C, SPI, UART, and GPIO. Added to this is the ARM TrustZone security protocols and encryption and the ARM on-chip interconnect bus. This system is pre-designed and configurable in terms of the number of cores, memory, and additional interfaces that may be added to the system, enabling designers to be able to scale their architectures according to their application needs. Additional driver modules can be easily added by the design team to handle Wi-Fi and other radio-based communications schemes, as well as protocols to handle a variety of interfaces to IoT sensing devices.

PROTOTYPE–TO-PRODUCTION

A second strategy is to make use of hardware prototyping systems both to design the system and to enter the market. In this strategy, multiple implementation methodologies are performed in parallel from the same CAD databases. For quick-turn prototype systems, the overall system can is partitioned into multiple ASSPs and FPGAs that are placed on standardized daughter boards with existing sensors and communications protocols already implemented. One benefit of this strategy is risk mitigation for the software/hardware interfaces. The prototype system enables the co-design of software and hardware using real IoT sensing devices and communications hardware. The prototype can be used to gather vital information from early users about how the system will really be used and the types of data traffic that can be expected from real-life use scenarios.

Multiple spins of the prototype can be made to iterate the design with customers until the right functionality is completed, at which point the new revisions to the system can be merged into one custom IoT SoC for the production versions of the system. The production versions are cost-optimized versions that take longer to complete and manufacture but run faster, use less power, and cost less than the prototype systems. Once tested, the production systems can replace prototype systems in the field with the same functionality but better overall performance and cost.

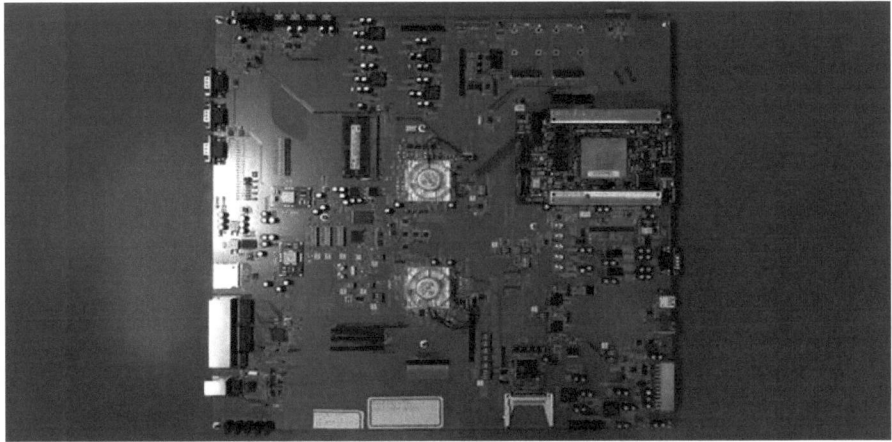

Open-Silicon IoT FPGA prototype board. Image courtesy of Open-Silicon.

TURNKEY ASIC COMPANIES

The third strategy is to make use of turnkey ASIC companies like Open-Silicon who have experienced design teams that have successfully completed hundreds of designs for their customers. Open-Silicon is unique in this space as they can handle all of the implementation activities of the system, including architectural trade-offs, system partitioning, hardware/software trade-offs, IC and board design, interfacing to manufacturing, assembly and test, and doing hardware bring-up of ICs, boards, and the final system.

Custom SoCs for IoT: Simplified

Open-Silicon already uses platform-based design and design prototyping. This means that complex SoC designs can be implemented rapidly first as FPGA prototypes on custom daughter boards, which enable the system designers to start debugging their applications on live hardware, while Open-Silicon executes on the cost-reduced SoC version of the design.

The next sections of this book focus on an IoT case study and how Open-Silicon implemented a custom IoT SoC using their Spec2Chip methodology and flow.

PART 2 - SPECIFICATION TO CHIP/BOOT

The rest of this book covers the steps used to implement, fabricate, and test a custom IoT SoC using the Open-Silicon Spec2Chip process. The idea is to cover the essential areas of SoC design following a platform-based design methodology. It should be noted that designers have complete control and choice over how many of these steps they do themselves vs. outsourcing them to a company like Open-Silicon. The steps outlined are used regardless of who is doing the actual work. In addition to covering the conceptual steps, there will be illustrations of their use in a case study done by Open-Silicon that followed this process.

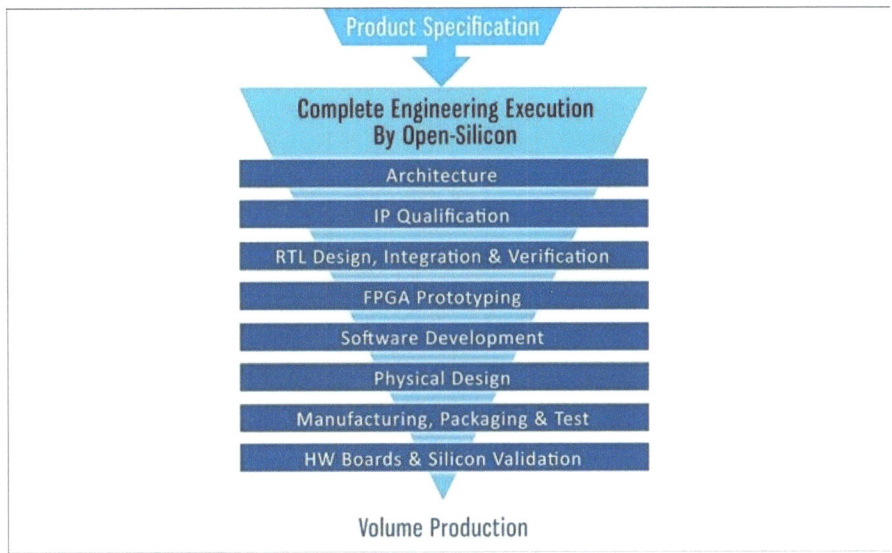

Spec2Chip design process. Image courtesy of Open-Silicon.

OVERVIEW OF SYSTEM DESIGN PROCESS

PRODUCT SPECIFICATION

The first step of any system is the development of a good system specification. Open-Silicon provides many customers with architectural analysis, hardware/software partitioning, and specification development to off-load the customer's internal teams. As part of their Spec2Chip methodology, Open-Silicon has developed a comprehensive system specification checklist, which is reviewed with their customers before any design work begins. The checklist covers the system-level functional and technical requirements including everything about the system's inputs, outputs, and processing functions, as well as how the system fits into and communicates with the larger IoT ecosystem.

The system specification includes documentation of the input data types and communications protocols used, a list of data transformation functions to be run and a list of output data types and protocols that will be used to communicate results. Also listed will be requirements for data communications bandwidth and latency between the device and the rest of the IoT ecosystem. Device boot up, control, and security are also documented along with proposed methods for delivering device updates once it is up and running in the field.

The system specification also includes information about the environment in which the device is to be deployed (temperature, moisture, exposure, etc.) and requirements for power, packaging form factors, installation, and maintenance capabilities. Lastly, the system specification will also document business goals of the system, including targeted market segments, development and recurring build-out and deployment budgets, expected pricing and profit margins, market entry strategy, and timelines.

SYSTEM ARCHITECTURE & PARTITIONING

Once the system specification is understood, Open-Silicon works with the customer to analyze and determine how to partition the requested functionality between hardware and software while considering performance (processing and latency), power (active and standby), cost (NRE and recurring – overall BoM), time-to-market, schedule, and risk. All of this translates into a myriad of trade-offs for the numbers and types used of standard vs. custom chips, intellectual property (IP) blocks, custom IC packaging, custom boards, etc. At this stage, decisions are made about how the hardware portion of the system will be partitioned amongst one or more chips and boards with decisions being made about the numbers and types of processing units, on-chip memory, communications interfaces, and security for each chip. Package selection and board decisions are key steps at this stage as both can have significant cost and performance implications for the overall system.

Custom SoCs for IoT: Simplified

As mentioned earlier, to reduce complexity, designers review and trade off possible pre-designed platforms that could be used to implement the base functionality of the design while also considering the availability of compatible software development platforms and IP blocks, and ease of configurability for design differentiation and specifications that are still in flux. Open-Silicon provides an IoT platform for just this purpose.

Open-Silicon's IoT platform has been designed to use the ARM Cortex IoT architecture in conjunction with a pre-designed IoT reference board. The reference board enables the IoT system first to be implemented with standalone ARM cores, memories, hardware device drivers, customized design logic (ala FPGAs), standalone sensors, and a pre-verified software stack for the subsystem. Software designers can write and debug their applications and driver code against the reference software stack and reference hardware using real-world system data while the SoC is being designed and implemented. Reference boards are typically outfitted with multiple different capabilities (both in terms of the hardware interfaces on the board, as well as drivers for the software stack). During the initial architectural phase, designers can use these multiple capabilities to hone in on the exact functionality they will need for their custom IoT SoC.

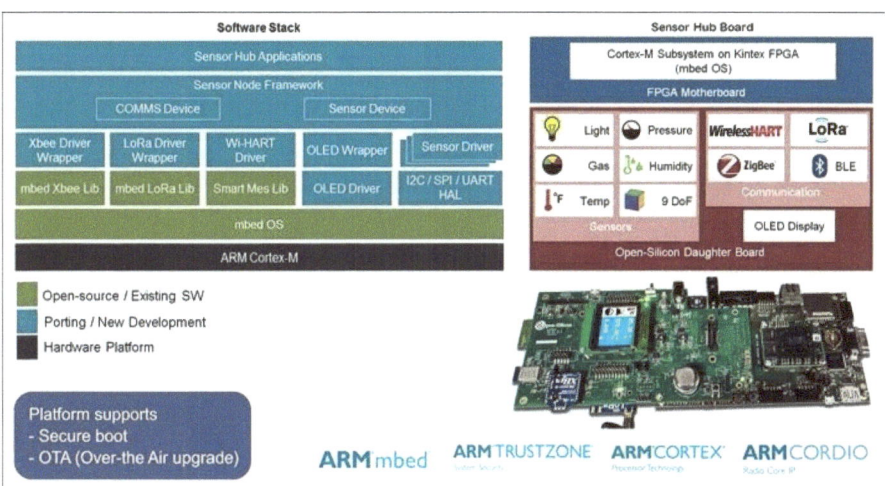

IoT edge platform with ARM-based mbed RTOS. Image courtesy of Open-Silicon.

SoC Technology and Package Selection

During system and hardware partitioning, the SoC team must also determine the silicon process and associated packaging technology that will be used for the system's custom ICs. Ideally, the least expensive process and packaging technology would be desired that meets the minimum requirements of the overall system. Trade-offs are often more complex than you might think as chip die size and expected yields for various foundries vary greatly. The team must also ensure that the chosen foundry can deliver the volume of chips required by the market being addressed.

Chip performance, IP block availability, the number of chip pins required, final packaged form factors, expected thermal environments, and many other factors such as the expected volume of chips to be manufactured, all play an important role in deciding which silicon process technology and package are the best choices for your overall system. Additionally, plans must be made for how and where the finished parts will be assembled and tested, especially if your IoT system is part of a safety-critical application. In these cases, there may be specific quality certifications needed before the system can be sold. Open-Silicon acts as a strong value-chain-aggregator with many foundry, packaging, and testing facility partners and can help designers make technology and packaging choices that are the right fit for the right price based on the requirements of the IoT system being built.

Open-Silicon has experience dealing with many types of packaging solutions and can help designers pick the right solution for their unique needs. Some of the packaging types with which Open-Silicon has experience include:

- Wafer-level chip scale packaging
- Low-cost lead-frame packages
- Full range of Ball Grid Array (BGA) packages
- High-performance flip chip packages
- Organic and ceramic packaging
- Multi-chip packages, including both stacked and side-by-side die configurations
- System-in-Package, where several die are placed on a silicon/organic interposer and this interposer is mounted on a package substrate for connectivity to package pins/balls

Open-Silicon also has experience with signal and power integrity modeling of packages and printed circuit boards (PCBs), including analysis of high-performance interface designs with greater than 28Gbps SerDes, high-speed DDR2/3/4, PCIe Gen3/4, WiMAX, etc.

SoC IP Selection / Qualification (Make vs. Buy)

IP blocks can be thought of as self-contained building blocks that implement specific functions. Sometimes these blocks are simply necessary functions that need to be in the system but are not necessarily differentiating to the design. In these cases, it may make sense for the design team to look to outside sources for the block instead of spending precious in-house design resources to create the block from scratch. Regardless of whether you choose to make or buy these blocks, it should be noted that the specification for these blocks must be completed before either the make or buy decision can be completed.

Additionally, the design team should also be aware of the following different views that will need to be either created or purchased for a given IP block. Note that in some cases IP blocks may be "soft," meaning that they are meant to be retargeted to a given process through logic synthesis, and some may be "hard," meaning they are already laid out to a specific target process and are no longer configurable. If purchasing the IP, any missing views will need to be created by the SoC design team.

Soft IP—Synthesizable RTL (register transfer language) for the block, including any synthesis constraint files

Hard IP—
- Layout abstract with pins and feedthroughs for place & route
- GDSII layout – full layer detailed layout that matches the place & route abstract
- Gate-level netlist (usually in Verilog format but could also be VHDL) used for fault grading, gate level simulations and mapping to FPGA logic for prototyping purposes
- Static Timing Analysis (STA) timing model (input loads, minimum pulse width, setup, and hold constraints, input slew, and output

load dependent timing paths, output drive capabilities - sometimes the netlist will have to suffice) generated from measured data of actual fabricated blocks

- Optional – some IP blocks may have packaged parts for the IP that can be used for in-circuit emulators, and daughter boards that can be attached to hardware emulators

Common IP Views for Both Soft and Hard IP—

- High-level model (Sometimes called a compact model used for system- and SoC-level logic verification)
- Verification test suites (e.g., test benches and simulation vectors used to test the IP block)
- STA constraints file (clock definitions, false paths, and mode conditions)
- Documentation on DFT methodology to be used for the block
- UPF power definition files and test benches for various power modes

A key consideration when purchasing IP blocks is to understand their specifications and assumptions of use. These should be well documented by the IP block supplier. Consideration should also be given as to whether the IP block has been fabricated in the technology process targeted for your IoT SoC. Having a block that has already been proven in silicon goes a long way towards mitigating risk, especially if the IP block supplier can provide references for its users and measured performance data from actual production usage.

In many cases, custom IP blocks are designed with functionality that differentiates the IoT SoC design from competing products. Customizing this functionality in hardware is an excellent way to protect the IP from being copied by competitors. These IP blocks are typically designed from scratch, although they too may make use of third-party IP blocks within them. As already mentioned, the design of these IP starts with a specification, followed by either a high-level simulation model or an RTL description of the block. A high-level behavioral model is helpful when trying to iterate to a proper functional specification for the block. The model can be used to debug the IP block against the rest of the system until the function is refined, at which point RTL for the block can be written to match the behavioral model functionality. Behavioral models can be written in programming languages like C, C++, or python. Care should be taken to use a language that is compatible with the rest of your CAD system and RTL-level simulation environments.

Other considerations for make vs. buy of IP blocks are the cost model for the use of third-party IP vs. the time and expertise required to build the IP from scratch. Remember, the cost to design your own IP includes not only the design of the IP logic but also the creation and verification of all the views just mentioned. In the cost calculus, you must also consider the volume of units you expect to manufacture over the life of the product. Third-party IP costs will usually contain, in addition to an NRE component, a recurring cost-per-unit component, whereas IP designed internally only has the NRE cost component.

As already mentioned, Open-Silicon has relationships with many third-party IP suppliers and has already done qualification work on many of the available IPs across multiple different foundry processes. Using Open-Silicon's services to help find and select IP blocks can greatly decrease the time and effort required to acquire the right blocks that will match your functionality needs while also mitigating risk by ensuring that block's compatibility with your targeted foundry process.

DESIGN AND VERIFICATION METHODOLOGIES
Before design work starts, project management must agree on the design and verification methodologies that will be used across the entire project. Areas that need to be agreed upon include:

- SoC power management scheme
- SoC design-for-test scheme
- SoC design-for-manufacturing methodologies
- Board and SoC design verification methodologies
- Silicon validation methodology

A brief description of each of these areas follows.

SoC Power Management Scheme

Power management can be critical for certain IoT applications, especially for IoT edge devices that may be operating on battery power. System designers typically set up a power management scheme appropriate to the SoC application during the micro-architecture design stage. This power management scheme can be captured in the IEEE 1801 UPF standard format.

UPF stands for Unified Power Format and can be used to capture the power intent of a design. Designers use UPF to specify multiple power domains within the SoC, their voltage levels, and which IP blocks are tied to each domain. Designers can also specify whether any given block can be dynamically transitioned into one of multiple lower power consumption states. The format enables designers to specify the signals used to control block power up/down sequences and what to do with memory contents of blocks that are being powered down.

Additionally, the format can also carry information for how signals that transverse power domains should be shifted at the domain interfaces. UPF is now supported in many different RTL-level simulators and can be used to generate functional test stimuli to exercise the assembled SoC power domains as specified in the UPF. This helps designers find and fix power related control logic problems before the actual IC is fabricated.

Custom SoCs for IoT: Simplified

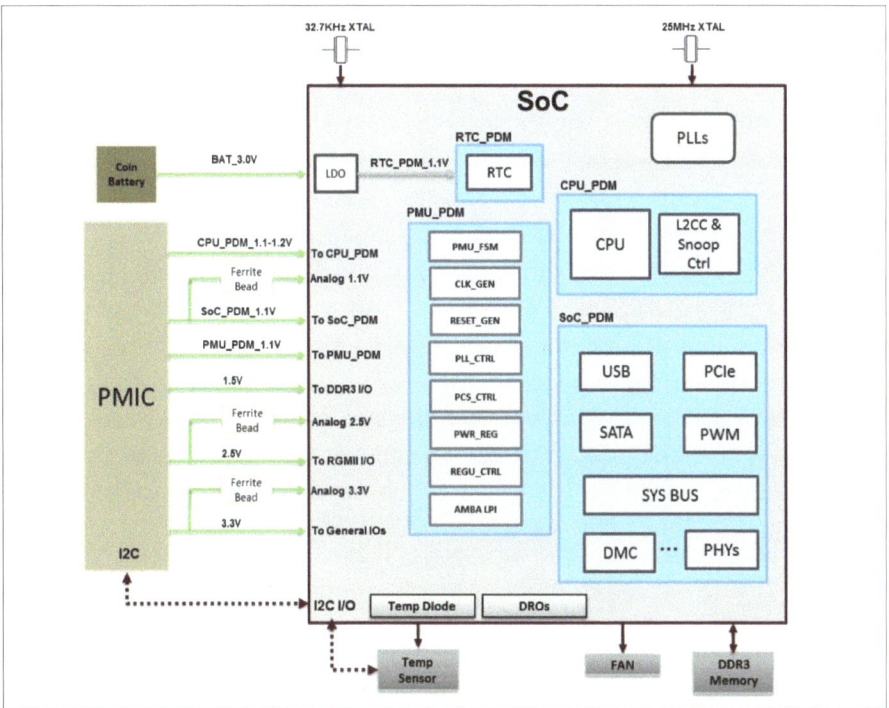

Example power management scheme with four power domains, six clock domains, and five power modes. Image courtesy of Open-Silicon.

SoC Design-For-Test Scheme

Like power domains, system designers need to have a comprehensive strategy in place for how the IoT SoC will be tested at wafer probe, after packaging, at silicon validation, and again once the SoC has been placed on the system board. There are several good generic strategies for incorporating testability into an SoC circuit. The base premise of design-for-test (DFT) is what is known as controllability and observability. The idea of DFT is to know if any given node in a circuit has a fault caused during manufacturing. To do this, the tester must be able to control that node's logic level from the device pins and also be able to observe the resultant logic level changes, again from the device pins, to ensure that the node is not stuck high or low. Controllability and observability is quite challenging for large circuits where internal nodes can be buried hundreds of levels deep in logic. Designers have learned that they can divide-and-conquer this problem by carefully designing the blocks of logic to be testable as a unit, and then having a strategy to be able to access each testable unit through a series of logic protocols that are also testable.

This same strategy is also applicable at the board level where standards such as JTAG 1149.1 can be used. The JTAG architecture enables data to be scanned into outputs of driving ICs and then clocked and captured at the inputs of driven ICs. This allows signals to propagate across the board interconnect between the ICs to ensure that the board interconnects are in fact functioning as designed. Additionally, JTAG can also be used as a port into the SoC that enables test patterns (inputs and expected outputs) to be shifted into specific SoC IP block registers for testing. Further, this same JTAG port can also be used to program the internal flash memories in SoCs while debugging during silicon validation and application software testing with the SoC.

DFT is a field of expertise in its own right and can be critical to the success of the IoT SoC, both in terms of the quality of the products being released from manufacturing and the overall costs of the product. If product or manufacturing defects are not caught before the product is shipped, it can be very costly to update the SoCs with fixes once they have been deployed in the field. DFT experts refer to this as the amount of coverage your test suite has for your design. You will also hear this referred to as the amount of fault coverage, although that term is now highly overloaded as faults are no longer simple stuck-at 0 or stuck-at 1 failure mechanisms.

Even when fault coverage is good, there is an art to coming up with the minimum number of test vectors that can be used to achieve the desired fault coverage. This is important because each additional test vector adds to the time that the SoC resides on the automatic test equipment. Tester time is a real cost associated with each unit shipped and thus contributes directly to the unit cost of each IoT device. Open-Silicon is very experienced at creating cost-efficient high fault coverage test vectors. More importantly, they understand the test problem from a design point of view and can help designers architect and implement a good DFT strategy that will also tie into the hardware/software prototyping capabilities they have with their IoT platform.

Custom SoCs for IoT: Simplified

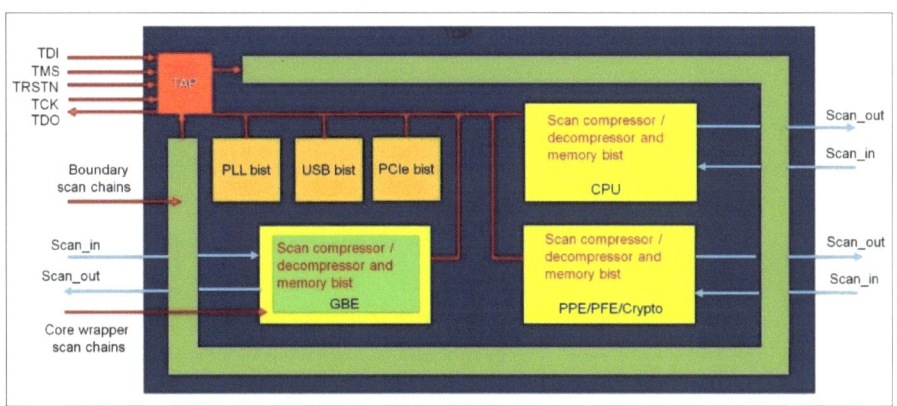

Example of SoC DFT strategy. Image courtesy of Open-Silicon.

SoC Design-For-Manufacturing Methodologies

In addition to ensuring the IoT SoC is functioning properly with the correct timing and power modes, designers must also think about how to ensure that the design will be robust to shifts in manufacturing. All manufacturing processes are somewhat imprecise. If engineered well, these processes are expected to operate within specified tolerances for key metrics. The combined effect of these tolerances on key metrics defines what is known as a good process window. Designs with features that fall within the process window will have high functional and parametric yields. Designs with features that fall at the edge of or outside of the process window will incur yield loss. Yield loss resulting in incorrect functionality can mean higher unit costs, depending on the business agreement in place with the foundry manufacturing the devices. Yield loss resulting in devices that function correctly but don't have the specified performance can also mean higher unit costs or worse, not having sufficient quantities of devices to fulfill unit orders. In some cases, lower performing devices can be binned or separated out at test time and those units may be able to be sold at lower premiums.

Many strategies can be employed to ensure robust and manufacturable designs. Some of these strategies will be enforced as mandatory design rules by the foundries. An example of this is the correct use of area-fill shapes in layout to ensure good control of wafer planarity as the design is manufactured. Area-fill also ensures good critical dimension (CD) control as both lithography and etch processes are load (or pattern) dependent. CD control is important as it dictates how well transistors match each other across the SoC die.

Other strategies will be considered good design practices but may not be enforced as mandatory design rules by the foundry. A good example of this might be the doubling of vias during routing. This will vary from foundry to foundry and from process node to process node. In some more aggressive process nodes, via doubling may be mandatory. In larger more mature nodes it may not. Regardless, it is a good practice if there is enough space to accommodate the extra vias as it helps reduce power and it ensures better lifetime reliability of the circuits.

Some examples of typical DFM treatments that may be done to a design include the following:

- Area-fill to ensure good planarity through density control – highly dependent upon design content (processors, memory, random logic, etc.) and SoC floorplan
- Area slotting or cheesing (too much material can be just as bad as too little)
- Via treatments (via doubling, bar vias, larger via pad enclosures)
- Restrictions on wrong-way routing (e.g., shapes only in preferred directions)
- Width and pattern-dependent spacing rules, especially around ends of lines
- Pattern restrictions (forbidden pitches, T-junctions, coloring conflicts)
- Stress-related patterning and overlap issues (important for front-end-of-line layers used for building transistors)
- Power bus width rules to ensure reliability (for electromigration mitigation and to alleviate IR drop, which can affect power rail voltage levels)

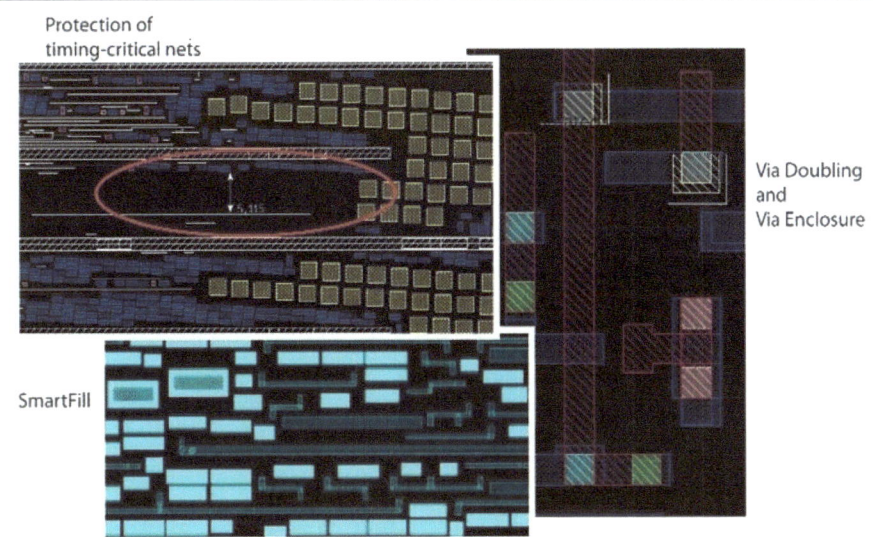

Calibre YieldEnhancer automatically modifies the layout to improve robustness of the design and yield. Modifications are inserted back into the design database.

DFM methodologies: smart metal fill, via doubling, via enclosures, and special spacing rules for critical nets. Image courtesy of Mentor, a Siemens Business.

In general, as you move from larger more mature process nodes to smaller more aggressive process nodes (e.g., 90nm to 28nm), you will find that many of the recommended DFM treatments of the older nodes become mandatory treatments in the more aggressive nodes. There is a significant breakpoint in process nodes where DFM strategies start to become very invasive in how the design is implemented. This breakpoint usually happens in the transition between 28nm and smaller process nodes. It is at nodes less than 28nm where multi-patterning lithography is employed to make the smaller geometries.

Multi-patterning is a manufacturing strategy that uses two or more masks to pattern one layer. By splitting the masks, each mask can be patterned at relaxed pitches. However, care must be taken with respect to mask alignment during processing to ensure proper patterning. The use of multi-patterning makes the SoC layout tasks much more difficult for both the CAD tools and the designers. If your IoT SoC requires using sub-28nm process nodes, you will likely want to enlist the aid of a design house, like Open-Silicon, who has experience doing these types of layouts as it requires much more experience and planning to accommodate the restrictions of multi-patterning.

SoC and Board Design Verification Methodologies
Virtual Prototyping Platform
System designers begin initial system simulation using high-level behavioral models for large architectural blocks of the system. If a pre-defined platform is being used, the platform supplier will many times have these behavioral models already written and verified against implementations in silicon. Some of the platform providers even have their own cycle-accurate simulators as part of their system design tools. This is referred to as a Virtual Prototyping Platform.

SoC Verification
It should be noted that system and SoC validation can take large amounts of resources both in terms of engineers and computer hardware. Designers will many times submit extremely long simulations that are distributed across thousands of servers in internal compute farms or externally on the cloud. Even with large compute farms, only a relatively short period of 'real time' can be captured in a simulation. Because of this, much engineering time is spent trying to intelligently set up simulation stimulus and expected results that can be used to pre-set the design into given states. Once a given state is set, simulation proceeds and simulation outputs are captured and compared against expected results. The use of DFT capabilities, like a JTAG tap ports, is extremely useful for this type of simulation debug.

Depending on the complexity of the system, verification can be one of the longest and most expensive design steps. Many designers will tell you that system and SoC verification are never complete. We simply can't run long enough simulations for that. SoC verification typically gets broken down into different functional groups of tests such as:

- Use-case validation
- Boot-up modes
- Performance analysis
- Regression and stress testing

Custom SoCs for IoT: Simplified

As mentioned earlier, the advantage of using pre-defined and pre-verified platforms is that these platforms and their IP blocks have already been debugged and typically come with their own test benches, stimulus, and expected results, sometimes referred to as verification IP. This can save the SoC verification engineers weeks of effort writing and debugging test vectors for the various IP blocks of the SoC. It can also ensure adequate functional coverage of the platform.

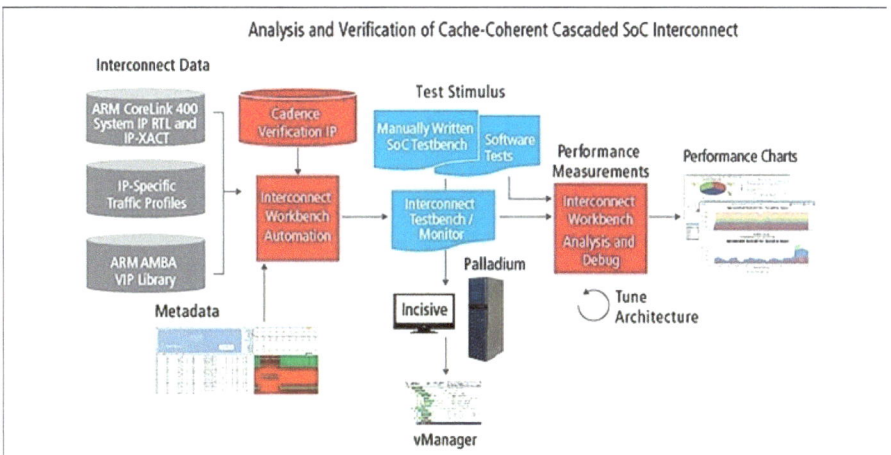

SoC verification using verification IP from standard platforms. Image courtesy of Cadence Design Systems.

Board-level Verification

Board verification applies to each of the multiple different boards that get created during the design process. This includes reference boards, FPGA prototyping boards, silicon validation boards, system prototype boards, and final system production boards.

For all these boards, there are several common verification steps that need to be carefully followed to ensure the boards will work correctly for their given purpose. It is important that the design teams agree upon the verification methods that will be used for the various boards as each board must be designed in such a way as to enable their specific purpose.

Some of the key verification steps include, but are not limited to:
- Interconnect functionality
- Power Integrity checking
- Signal Integrity checking

- Interface functionality
- Integration testing

Board Interconnect Functionality. As mentioned previously, board traces are typically tested by using JTAG functionality or test logic on the board. The idea is to isolate each individual trace on the board and ensure that signals are propagating correctly from source to sink of the trace. This may involve more complicated testing for high-speed signals that require certain matched impedances where special care must be taken to de-embed testing equipment so that measurements are accurate.

Power and Signal Integrity. Power integrity checks ensure that the board has the appropriate trace widths and power source regulator components to supply chips on the board with proper voltage levels and required currents. This includes making sure the power rail voltage levels are maintained under stressed loading and performance conditions with worst-case switching scenarios that can tend to degrade power rail voltage levels.

Power integrity checks are also made on the board to make sure that the proper amount of decoupling capacitance is being used (either on-chip or on-board) to ensure noise on the power lines is not affecting chip functionality or performance levels.

Depending on noise conditions, board designers may need to work jointly with chip designers to ensure an adequate number of power and ground pins are available to isolate noise-sensitive signals from high speed interconnects on the board. Checks are also made to ensure high-speed signals are adequately shielded to avoid crosstalk between board signal traces.

Before designing of the board, pre-layout signal integrity analysis is performed on high-speed digital interfaces. The results of the analysis will provide rules for layout on what should be the trace length, width, distance between traces, crosstalk between the traces, maximum speed the traces can reach, etc. Based on these design rules, the layout for the high-speed section of the board is laid out. After the layout is completed, post-layout signal integrity analysis is performed, which will provide results that will be close to the results measured on the physical board.

Power integrity analysis is performed to analyze proper power distribution for different power rails on the board. Any anomalies in these power rails will directly impact the performance of the chip to which the rail is providing power. The selection and placement of decoupling capacitors is very critical for any high-speed board. Using power integrity analysis, the proper value and efficient placement of decoupling capacitors is determined, which will also help in reducing the noise entering the chip.

Using signal integrity and power integrity analysis helps to design an efficient and first- time success board.

Interface Functionality. Once board signal traces are confirmed as good, various interfaces are tested next. The main point of these tests is to check multiple pins on each die and the board that must work together with a specified interface protocol. Each supported interface must be tested to ensure the combinations of die, pins, and interconnects work properly for the protocols required.

For system boards (as opposed to silicon validation boards), the board designers may also run interface checks while varying voltage sourced to the board. The same can be true for varying ambient temperature over the board while running interface tests. These variations in voltage and temperature ensure that the chip works in various operating conditions. A production board may have a voltage regulator providing voltage to the chip, which may have some variation compared to the recommended voltage. These variations will be validated on the system board by varying the supply voltage and making sure that the chip works flawlessly under these voltage variations. Similarly, the product in which the chip is used may be deployed in a hot climate, like the desert where the temperatures may reach 50°C, or in the mountains where the temperature may be well below 0°C. The chip under test on the system board is subjected to forced hot and cold air so that it mimics the actual temperature in adverse weather conditions and tested for intended functions. These steps will make sure that the chip is qualified to be placed in products that will be used by consumers around the world in varied conditions.

Integration Testing. Finally, once the physical aspects of the board have been verified, designers will then start integration testing of the board with peripherals and application software. The desire here is to check for overall system functionality and performance.

Depending on the type of board being verified, there will be different criteria for pass, failure, or re-work. As an example, reference and demo boards are not necessarily expected to run at final expected system speeds. Designers, however, must verify that the board design does not somehow negate or prevent the system from working at the slower speeds.

SoC Design and Implementation

Once the system architecture and partitioning decisions have been made, many different tasks start in parallel to design and implement the system. A high-level list of parallel activities includes:

- Hardware/software co-design with an FPGA prototype platform
- Application software development
- SoC micro-architectural design
- SoC power management methodology
- SoC design-for-test (DFT) methodology
- SoC design-for-manufacturing (DFM) methodology
- SoC logical (RTL) and physical design (layout)
  Board design
- FPGA prototyping board
- SoC validation/bring-up board
- Production board

It should be noted that engineers tend to specialize in one or more of these areas. A quick description of engineering skill sets, followed by a short description of each of the main design activities, follows.

Design Team

You may see engineers in groups as follows:

- System Designers (system design and verification)
- SoC Logic Designers (code RTL, synthesize and time the design)
- SoC Verification Engineers (create simulation test benches that exercise the SoC functionality)

- Software Engineers (write application code for the system, silicon validation and bring-up)
- Board Designers (board design and physical board layout for FPGA prototype, silicon validation, and production boards, silicon validation and bring-up)
- SoC Physical Designers (floorplanning and hierarchical layout using automatic place & route software, DFM, complete final logic and layout optimizations to meet timing requirements)
- SoC Test Engineers (develop test patterns for automatic test equipment, ensure high fault coverage with the shortest possible tester time)
- SoC Product Engineers (perform silicon validation and bring-up, work on design changes for yield improvements once the design is in production)
- Board / System Test Engineers (develop test strategies for board test and final system bring-up on production boards)

Very rarely do you find a single engineer who takes a design through all of these steps. For smaller to medium size companies, it tends to be more practical to have the system designers and perhaps an expert or two who cover semiconductor and board manufacturing and then outsource the rest of the jobs to design houses, like Open-Silicon, that have large numbers of these various types of engineers. The costs of these engineers are amortized by the design houses over multiple system customers' work.

HARDWARE/SOFTWARE CO-DESIGN WITH FPGA PROTOTYPING PLATFORM

One of the trickiest parts of designing a custom IoT SoC is the co-design of the hardware and software components. Picking a well-established platform that has support for compilers, debuggers, real-time operating systems, device drivers, and security is a must. Having to iterate a custom IC is an expensive proposition both in terms of additional NRE and, more importantly, the opportunity cost of missing market windows. Given this, it is essential that you have a way to debug the application code against the real device drivers and interfaces that will be used by the system before the SoC is fabricated.

As previously discussed, reference platforms are typically outfitted with multiple different capabilities, both in terms of hardware interfaces and drivers in the software stack. Once the architectural decisions are made as to which of these capabilities are required for the IoT system, the designers have two choices. They can either build a new custom FPGA prototyping platform that is designed to use only the capabilities that will be required for the production IoT system, or they can use off-the-shelf FPGA prototyping platforms from companies like proDesign, Synopsys, Dini, S2C, and Cadence Design Systems. In the latter case, the customer uses the supplied FPGA platform and builds peripheral daughter cards that plug into the FPGA system.

The first approach will be expensive and time-consuming due to custom tool and software development requirements, but may be needed if off-the-shelf platforms do not meet design needs. The second option is also expensive, but development will be faster, and you can expect tool support to be provided by the supplying vendor. Vendor-supplied platforms will also be already pre-verified and known good.

As the name infers, the FPGA prototyping board uses FPGAs to enable custom logic to be integrated with the already debugged and tested IoT platform IP. It is this board that will be used to co-develop software against a functional system while the SoC is being implemented and manufactured. System architects will also use the FPGA prototyping board to build and verify system tests for the final production system.

The FPGA prototyping board is specially designed to interconnect to a host of standard IC interfaces, hardware emulators, and in-circuit emulation (ICE) capabilities. Setting up a system prototype enables IoT system designers, SoC hardware designers, and software applications designers to exercise their logic against real-world environments with sensors, actuators, and communications protocols. The prototype system is cycle-accurate and runs multiple orders of magnitude faster than software simulations, enabling 'real time' interactions of the prototype system with the 'real world.'

As previously covered, Open-Silicon's IoT platform has been designed to use the ARM Cortex IoT architecture in conjunction with a pre-designed IoT reference board. Application software designers start writing and debugging their code against the pre-verified IoT software stack that comes with the platform. As major blocks of code are completed, they can be tested first against the reference design. This proceeds in parallel to SoC designers implementing the hardware micro-architecture in RTL. As the major blocks of the hardware are designed, they can be implemented from the RTL as logic in the reference board's FPGAs against which the software can be tested.

Because Open-Silicon has architected their IoT platform to use this design and test methodology from the start, the source for both the system prototype and the final SoC come from the same CAD database, and thus problems that are found and fixed in the system prototype can be easily reflected into the final SoC. This saves the design teams large amounts of time and provides design integrity between the prototype and final products.

APPLICATION SOFTWARE DEVELOPMENT

Open-Silicon's IoT platform provides a software development environment that is pre-configured to work with the ARM Cortex-M IoT architecture. When using the Open-Silicon IoT platform, application software developers can focus on their application code while the rest of the software stack, including the ARM mbed operating system and device drivers, are already available as part of the platform. The platform also supports a secure boot environment for the device, along with facilities to enable over-the-air updates of application code, in case applications software needs to be revised or updated. IoT applications software developers have the software stack available as linked libraries and APIs that can be called to access device drivers and communications links provided by the system. This saves the application developers time as they can stay at the applications level and let the platform take care of software drivers for specific hardware interfaces and security protocols.

SOC/MICRO-ARCHITECTURE PARTITIONING

As previously discussed, an IoT SoC can be relatively simple or very complex. IoT gateways that deal with a fusion of complex and varied data often require cooperation between heterogeneous specialized processing units. If the IoT device is making real-time decisions, it may also have to incorporate coherent memory access schemes to keep the cores in sync with each other. Once functions have been assigned to individual SoCs, the next step is to decide upon a micro-architecture for the IC.

Custom SoCs for IoT: Simplified

The easiest way to do this is to choose from existing platforms that have already been well designed and debugged. Open-Silicon's IoT platform makes use of the ARM Cortex-M IoT offering that allows for inclusion of one or more ARM processor cores working in conjunction with multiple on-chip memories, interface blocks, and other third-party or custom IPs that know how to interface to the standard ARM system interconnect architectures. Open-Silicon has relationships with IP suppliers such as Synopsys, Cadence, CEVA, S3, Rambus, TCI, and others and can help their customers to qualify and select a wide range of IPs, such as SerDes, analog/wireless, interfaces, processors, digital signal processors (DSPs), graphics processing units (GPUs), memories (SRAMs, ROMs, CAMS, etc.), standard cell libraries, and specialty I/Os. Because Open-Silicon works with these vendors on a regular basis, they know which IPs are qualified in various foundry processes and which ones have already been tested and debugged with their IoT platform.

At this design stage, Open-Silicon works with the system designers to partition the hardware functions amongst the processing units and IP blocks that will be used in the IoT SoC(s). Typically, a block diagram is constructed for each SoC, highlighting its major system blocks, memories and external interfaces along with the system interconnect bus or on-chip network that will be used to tie it all together.

Custom SoCs for IoT: Simplified

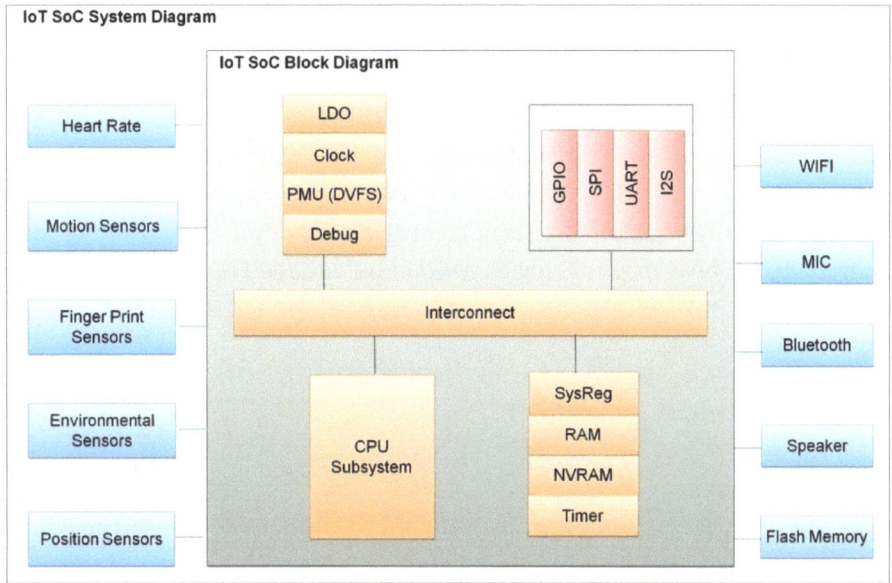

Example of IoT SoC system architecture. Image courtesy of Open-Silicon.

Since Open-Silicon works closely with ARM platforms, they use ARM's high-level system design tools to capture the designers' unique intent. This software lets Open-Silicon quickly assemble building blocks into their predefined IoT platform architecture using a simple-to-use graphical user interface. Because the system design software knows the base platform architecture, assembly of a unique system is as easy as specifying the types and number of processing units, memories and hardware interface protocols to be supported. The system makes assumptions about how these units will be connected through the platform interconnect scheme and builds a model of the system that can be simulated at a functional level using cycle-accurate logic simulators. Designers work in conjunction with Open-Silicon to make architectural trade-offs for SoC throughput, power modes and basic interactions with external devices through the specified interface protocols.

As the SoC architecture solidifies, the system design software can then be used to automatically generate RTL code for all the logic connections that will be used to tie the SoC system components together using the platform's interconnect scheme. In parallel to this, custom IP blocks can also be integrated into the system at the RTL level, and the system simulations can be repeated with more information using RTL level logic simulators.

It's also important at this stage to have a good understanding of strategies that will be employed for power management, DFT (including SoC bring-up and debug on the board), DFM, and hardware security as these methodologies tend to put constraints on all blocks of the SoC and system interconnect. These methodologies need to be designed in from the beginning and not added on as an afterthought at the system and RTL levels.

SoC RTL Design

Once the system has been partitioned into hardware and software, and the hardware has been partitioned into individual SoCs, the SoCs can then be represented by RTL models for the design blocks and hardened IP. The RTL provides the connectivity and Boolean logic between the blocks and can be used by RTL-based logic simulators to do cycle-accurate functional simulations. This is sometimes referred to as an RTL platform, meaning that the entire system has been captured at the RTL level and can be simulated.

SoC logic designers write RTL code that will be used to synthesize the logic that combines the IP blocks into the overall SoC. Much of the RTL may be automatically generated by the SoC platform design tools, but logic designers need to manually write RTL code for their customized IP blocks, the connections between those blocks, and the chosen platform's interconnect bus IP. Logic designers will also write RTL for logic that will control the various power modes supported by the SoC. They will also need to make the logical connections of the platform IP and its interface IP blocks to the appropriate chip I/O drivers and pads.

SoC Implementation

This section discusses the following implementation topics:
- Top-level SoC planning
- Block-level layout and verification
- Top-level SoC layout assembly

Top-level SoC Planning

Top-level design planning of the IoT SoC provides an initial layout template for the IoT SoC. Initial placements of IP blocks with known size and aspect ratios are made along with estimates of soft IP blocks that will be synthesized and laid out as the SoC implementation progresses. Special care must be made to ensure adequate power and ground can be routed to the various IP blocks based on their estimated power consumption needs and power domains. Estimates are also made for the number of metal layers that will be needed to make routing connections between the blocks based on the top-level block netlist. One of the key benefits of using third-party IP blocks is that quick synthesis runs can provide an estimate of gate count and area required for the soft IP blocks, which can then be used for both area and power consumption estimations which can then be used for block and power bus sizes while floorplanning.

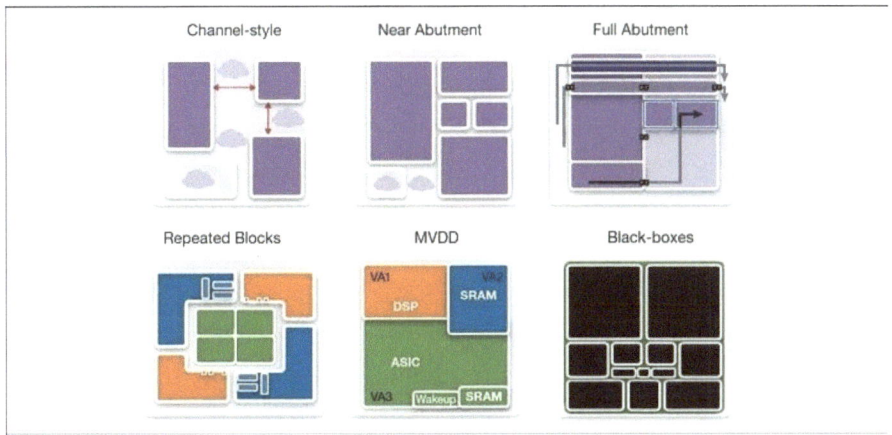

Top-level SoC planning methodologies. Image courtesy of Synopsys.

Block-level Layout and Verification

Once an initial floorplan is set, top-level power and signal routing starts and individual design blocks are farmed off to multiple logic and physical design teams to be implemented. Each team is given their target layout area and aspect ratio along with timing constraints and estimated pin locations for I/Os that cross their block boundary. They will also be given a restriction on the number of metal layers that can be used for their intra-block connections and where power bus connections are expected to come into the block.

Logic designers and physical designers iterate synthesizing and performing trial place and route runs to converge on the design within the given constraints. As the process proceeds, some teams will find that they need more area or a different aspect ratio and those requests get iterated up to the top-level chip layout specialists who try to steal from Peter to pay Paul. The same thing can happen as top-level designers find issues with inter-block level routing. These iterations can cause I/O pins of sub-blocks to have to move locations, which then causes another iteration of the layout at the sub-block.

In parallel to this effort, the design teams also have specialists that review the status of each block and checks the blocks for adherence to logical, timing, electrical, DRC (design rule checks), DFT, and DFM requirements. Priorities are established with logical, timing, and electrical coming first (e.g., if it doesn't function correctly or have the right timing and electrical behavior, it doesn't matter if it meets DRCs, DFT, or DFM requirements).

Tools used to check these steps include the following:

- Formal Verification
  Used to check RTL vs. gate-level netlist equivalency.

- Parasitic Extraction
  Used to determine parasitic capacitance and resistance for nets based on actual routing. Back annotated to static timing analysis and used to determine delays as the signal passes through logic gates and metal interconnect.

- Static Timing Analysis
  Confirms logic signals get to block outputs at required times given assumed input signal arrival times, timing modes, and timing variations due to process variations.

- Clock Tree Synthesis and Analysis
  Used to generate clock tree buffering to ensure clocks arrive within skew tolerances for the circuit. Also checks for proper synchronization of multiple clock domains.

- Signal Integrity and Noise Analysis
  Checks for logic and timing errors induced by signal crosstalk,

ground bounce, and power supply noise. Crosstalk can induce glitches on clock and reset lines that can cause incorrect function.

- Thermal Analysis
  Checks or hot-spots and excessive self-heating caused by fast switching circuits or circuits for which there is insufficient metal width traces to carry higher current loads.

- ERC - electrical rules checker
  Confirms logic connections follow appropriate protocols, such as fanout limits, correct logic used for transitions between voltage domains, etc.

- DRC – design rules checker
  Confirms all layout spacing and metal widths conform to manufacturing rules. These can be quite complex at more advanced process nodes.

- Power and Electromigration Analysis
  Confirms that the metal power buses are sufficiently wide so as to not have too much voltage drop across the block. Also looks for layout geometries that may be likely to have electromigration problems over the expected lifetime of the chip.

- ATPG – Automatic Test Pattern Generation / Fault Grading
  Generates test patterns that will be used by ATE (automatic test equipment) during wafer probe and die testing before and after device packaging. Low fault grading scores can result in logic changes needing to be made to improve testability.

- DFT-checker
  Analyzes logic to find signals that may not be accessible from the block pins for testing purposes. Also, checks that certain test access ports are correctly connected. These will be flagged, and additional logic will need to be added to improve testability.

- DFM-checker
  Analyzes the layout for layout patterns that may prove to be problematic for advanced lithography. Scores the layout for its robustness against random and system yield limiters.

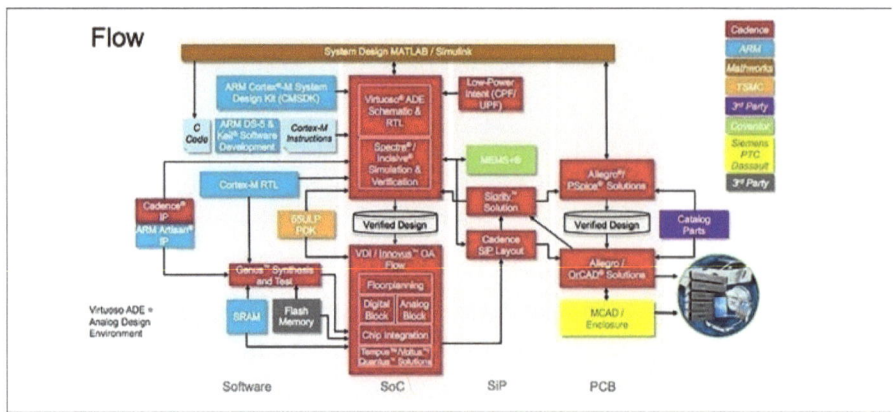

Electronic design automation (EDA) design flow. Image courtesy of Cadence Design Systems.

Top-level SoC Layout Assembly

As with the lower level blocks, designers synthesize the top-level inter-block connections and then place and route that logic and all other connections between the blocks including power and DFT connections. Analysis of this logic and routing at the top-level uses the same steps and tools as used at the lower levels. Most of these checks are done with the blocks being represented by simple black box models. As design blocks solidify, they are merged into the top-level floorplan to ensure assumptions made about their size, aspect ratio, I/O locations, timing, electrical specifications, etc. have not changed.

When all blocks have been populated at the top level, a final timing check will be run to ensure the overall timing of the IC is being met. STA is also run at the top level of the SoC to ensure timing budgets are being met. Depending on the STA tool being used, this may require very large full-chip runs, or it may be done hierarchically through various methods of automatically creating timing models from the lower-level blocks. Additionally, final DRC, DFT, and DFM checks are made to ensure that combinations of blocks at the top level have not caused new errors or conflicts.

It should be noted that for large SoCs with multiple processor cores, memories, and advanced network-on-chip architectures, that these final verification steps are non-trivial and usually require more than simple brute force methods to successfully tape out the design. Companies are again advised to take advantage of design houses, such as Open-Silicon, who have developed advanced design methodologies over the years based on their experience of putting hundreds of these types of designs together in the past.

TAPE OUT DATA PREPARATION

Final steps before tape out include DFM treatments for things such as metal fill, metal slotting and cheesing, and possible biasing of certain layers to accommodate for mask and foundry-specific processing steps. Final DRC checks are made to ensure that none of these steps has induced any new errors. These steps are done at the end of the design process as their implementation is dependent on macro effects on the die based on layout densities in various parts of the chip. Also added are alignment marks and chip seal rings that are needed for correct mask and wafer processing.

Finally, depending on the process node at which you plan to have the design manufactured, you may need to run a series of lithography process checks to ensure that design is manufacturable. These steps are run to establish a high confidence that all mask data preparation algorithms will be successful per the rules of the mask shop and foundry lithography rules.

These checks use a combination of design rules and lithography-based simulations to determine how the layout will look after the foundry lithography process. If the tools find the shapes as printed are too different from the shapes as drawn, those shapes will be flagged as needing specific treatment. Lithography experts will then examine the layout and decide how to best compensate for the flagged problems. These areas are typically referred to as lithography hotspots.

MASK DATA PREPARATION

After tape-out, the design data is sent to your chosen mask shop. Open-Silicon can help you decide which mask manufacturer may be right for you, given your choice of foundry and technology process node. The mask shop will prepare the finished design data for the demands of sub-wavelength lithography. This preparation includes application of resolution enhancement technology, like optical proximity correction (OPC). Depending on the fidelity required for your design and wafer processing node, additional treatments may be made to the data to ensure good mask quality is met. After lithography correction, the data is fractured into polygons and written out in one of any number of mask writer formats.

Mask costs are a function of the types of mask writing machines that need to be used to reach the required mask fidelity level and the amount of time that the individual mask resides on that equipment. Inspection of higher quality (e.g., more aggressive patterning) masks also requires the use of more expensive equipment, which will also affect the price of the masks. Lastly, the number of layers, or masks, also will be a key factor for how much of your NRE budget is going to building your masks. Each layer will be written on the appropriate equipment required to meet the critical dimensions of the geometries you are patterning on that layer. Some layers will need ebeam-based mask writing, while other masks may be able to use less expensive laser-based mask writing. Once masks are inspected and found to meet specifications, they are sent to the foundry for wafer processing.

WAFER MANUFACTURING

For those companies who have never had a custom IC manufactured, it is highly recommended to use a design house, such as Open-Silicon, to handle the interactions with the chosen manufacturing foundry. There is a comprehensive checklist that must be followed for tape out to a foundry, and valuable time can be lost while trying to navigate the paperwork and checklists. Open-Silicon has done this hundreds of times and has dedicated personnel whose job it is to ensure that all necessary files are included in the hand-off and that process steps, as defined by the specific foundry, are followed. Once the tape out is submitted, Open-Silicon will also track the progress of your designs through the fab while preparations are made with the selected OSAT (Outsourced Assembly and Test) provider.

In addition to tracking, Open-Silicon can also provide product engineering services that will work with the foundry to establish minimal test pattern configurations for wafer probe testing. If the design moves into full production volumes, Open-Silicon can also work with designers to look for any yield limiters to the design to help bring overall unit prices down.

ASSEMBLY AND TEST

The same advice is given for interfacing with OSATs. Open-Silicon has partnerships already established with several OSAT vendors. This means that they have already worked with and qualified these vendors and know their strengths and weaknesses. Open-Silicon can provide valuable inputs on which OSAT would be the best match for your design based on the type of packaging and testing your design requires.

One of the most valuables services provided by Open-Silicon is the ability to work with OSATs to define and refine the test patterns that will be used to test the packaged parts. The trade-off becomes one of finding the right set of test patterns that will adequately test the device based on the quality required by the application while keeping unit costs down. The longer units are on a tester, the more the test time per unit costs. Ideally, Open-Silicon will have already been working with the design team on a DFT strategy that will aid in this process. However, the real test environment is where the rubber meets the road. Having a team of solid product and test engineers to debug issues on the test floor when the real packaged parts arrive can be crucial to a successful launch of a product, especially one that is as complex as an IoT SoC.

BOARD DESIGN

In parallel to system prototyping and software development, board designers begin working on initial design and layout of several different boards. These include:

- FPGA prototyping board
- Silicon validation board
- Platform reference board
- System/production board

FPGA PROTOTYPING BOARD

Reference boards are kind of like a Swiss Army knife: open one up, and you find all sorts of gadgets in there, many of which you will never need or use. They are meant to cover a wide breadth of possible problems and solutions, and they are quite useful in that regard. However, once the system designers have had a chance to review the reference board, they usually decide to pare down the functionality to only the essentials they need for their end market. At that point, they design their own customized FPGA prototyping board that will eventually be used to help them do software/hardware co-design.

The good news is that several companies already provide FPGA prototyping platforms. Regardless, the idea is to set up the board in such a way that it can be used (in a non-production way) to interact with the real world in a similar way to what the final IoT SoC will do. The other good news is that if you are using a design house like Open-Silicon, who developed the platform reference board, you also have access to the people that really understand all of the features of the platform and how it is meant to work.

The same thing is true for the software stack. The designers will carve out and use only those components of the software stack that are needed for their actual design. Through a series of system tests with the FPGA prototyping board, they can not only debug the hardware/software interactions, but they can get an accurate feel for how much on-chip memory they will need to store the software stack and their application.

If designing a custom FPGA prototyping board, care must still be taken to design them correctly, using the verification steps already outlined. The reason for this is that you don't want the FPGA prototyping board to give you false information (good or bad). That means it must be designed carefully to not accidentally influence your SoC design due to poor board design practices. Again, working with a design house like Open-Silicon helps to mitigate these unwanted and sometimes unnoticed problems.

IoT edge platform reference board. Image courtesy of Open-Silicon.

SILICON VALIDATION BOARD

A silicon validation board is something very different from either a reference board or an FPGA prototyping board. Its sole purpose is to help the designers validate the manufactured SoC's functionality and performance. In this case, the validation board has several features that help product engineers do this task.

The first difference you will note on a validation board is that, unlike a reference board, the validation board provides test access to each pin of the SoC. A second difference is that the validation board can also sweep voltage levels in very fine increments. This capability is used during both functional testing as well as characterization of the SoC across process, voltage, temperature (PVT), and corners. The validation boards have provision for mounting test sockets. These test sockets house the chip that has arrived after manufacturing and they have a provision for removing and inserting new parts. During validation, a number of chips need to be tested on the same board across PVT corners. Since it is not feasible to solder and de-solder the chip on the board multiple times, a test socket is used, which will ease the validation process.

Validation boards will also have a master clock(s) that drives the SoC, and these clocks can be adjusted for different jitter and delay tolerances. The idea here is to not just test the SoC at the optimum conditions for which it was designed, but also to see how the SoC performs when the environment isn't so optimal.

Like the reference and FPGA prototyping boards, validation boards also require care in their design and engineering for the reasons mentioned earlier. In short, the board must be designed such that there are no false positives or negatives in the SoC testing resulting from inadvertent board parasitics, noise, or poorly controlled board voltages and currents.

Validation boards are also used to do failure analysis when there are system failures in the field. The boards have access to JTAG ports and other test pins on the SoC that allow them to set up conditions on the SoC as they were when the design failed, and from there they can work backward to understand the root cause of the failure and look for ways to fix the problems.

PLATFORM REFERENCE BOARD

A platform reference board is a board that contains an implementation of a given platform with processor cores, memories, and a host of peripherals that would be applicable to a given end market segment. In our case, this would be an IoT reference system. The base reference board typically contains the processors, memories, device drivers, and interconnect fabric that would be used in the custom SoC. Peripherals are typically added as daughter boards that can be connected to the main board, enabling the designers to customize the reference setup. The base boards are designed to include one or more FPGAs that can be programmed to emulate the designer's custom ASIC logic functionality.

In addition to the hardware, the platform provider also supplies a software stack with an operating system and software libraries that have been designed to work with the various drivers found in the reference design. In addition to the software stack, there is typically a software development environment that gives software developers the ability to compile code for the target processor architecture. Software developers also need debugging aids that enable them to step through their code and watch interactions between the hardware and software through windows into registers and I/O pins.

Custom SoCs for IoT: Simplified

The good news for users is that if you are using a platform reference board, someone else designed it and took care of all of the tough design issues ahead of time. You just get to use it!

SYSTEM/PRODUCTION BOARD

System boards are the boards that will be used for the actual system product. However, as with other parts of the system, even system boards have stages where prototype boards are first designed and implemented with the actual parts that will be used (including the IoT SoC). However, the prototype boards may be bigger and include more debug capabilities than the final boards that will be used for production volumes. These boards are meant to be used in small quantities in the field for system testing with real IoT data.

System boards are typically designed and built by board houses that do production runs. Their engineers work closely and early on with the IoT system designers during the architectural and partitioning phase to do trade-offs for board configurations, dimensions, number of layers, SoC pinouts, etc. This work is very important as it drives direct recurring costs for the end system and can have a large impact on both the SoC and the overall system performance. Board designers will also heavily use lessons learned from the demo and validation board efforts. Like the SoC designers, board designers capture schematics of the overall board and use IBIS (I/O buffer info simulation) models that are produced during the SoC validation phase to do simulations for signal and power integrity, as was done for the other boards.

Prototype boards are also used for manufacturing process learning. Depending on the type of board and the IC packages used, the board manufacturer will set up a specific board manufacturing process for the system. This process will be monitored and tweaked to reduce costs and improve yields as the system ships in production. Prototype boards are especially useful during the manufacturing process bring-up phase to try out different ideas that will eventually be used for the final board production runs.

Silicon Bring-up

Once the SoC has been manufactured, packaged, and tested, it is mounted on a silicon validation board. Silicon validation boards typically support the following four specific types of tests that are used for what is known as 'silicon bring-up':

- Functional pin tests
- Interface tests
- Integration tests
- SoC characterization

Functional Pin Tests

During silicon bring-up, each pin of the SoC is tested to ensure it gives the correct results for every function and mode of the device. This includes power and ground pins where measurements are taken to make sure correct voltage and current levels are maintained in each situation. Basic level of software code is written, known as bare metal code, which will run on the SoC, and toggles interface pins so that these are monitored on the board at various test points.

Interface Tests

Interface tests cover sets of pins that are used for each supported interface protocol. Each interface that is to be supported by the SoC must be tested to ensure all hardware pins associated with the protocol are working properly. In these cases, the pins are attached to actual peripherals either on the validation board or on daughter boards, and electrical characteristics are tracked during typical usage sessions that are defined by the interface. This includes both function and adherence to any performance specifications as required by the interface protocol.

Integration Tests

Integration tests are tests that are set up to check interactions between the interfaces and the software stack, device drivers, and the applications software. These checks are testing the functionality of the system using peripherals that are connected to the validation board. The tests are very similar, if not identical, to the kinds of tests that were first performed on the FPGA prototype board except that in this case, they are now using the real

SoC instead of the FPGAs and discrete components that were used on the FPGA prototype board.

SoC Characterization

Lastly, the same set of tests described above are also run across multiple voltages and temperatures and die from different process lots (e.g., fast, slow and typical), as provided by the silicon foundry to characterize the performance of the SoC. Characterization tests are essential to understanding if there is any system degradation when the SoC is used over the possible variations that may occur in the use environment or due to differences in processing variations from the foundry.

System designers typically qualify their designs for certain temperature and voltage ranges depending on their end-market application. Examples of this include commercial (0^0C to 85^0C), industrial (-40^0C to 85^0C), or automobile (-40^0C to 125^0C) temperature ranges. Designers will test that their designs function properly over these ranges with +/- 5% to 10% margins. They will also use the validation board to sweep these tests for under or over voltage conditions, again usually testing with +/- 5% to 10% margins outside the stated voltage range over which the product is to operate. Besides using the tests to generate documentation and datasheets for the IoT SoC, these types of test are also especially useful to determine if there are any margin issues related to setup and hold conditions in the SoC logic that may be triggered by extreme environmental conditions found in the field.

Production Board Assembly and Test

In addition to the board verification tests already discussed, board design houses will also repeat many of the same software/hardware functional and parametric tests that were run on the FPGA prototype board. The exception is that, like the silicon validation tests, the board manufacturers will also run tests across multiple different environmental conditions that the system is expected to encounter in the real world. In these cases, regression tests will be run on the board while varying environmental temperature, humidity, vibration and other mechanical and environmental stressors.

Spec2Chip Case Study

This last section covers a case study in which Open-Silicon used its Spec2Chip process to design an IoT gateway reference design for their customer. The design project's internal code name was Voledia and, in this case, the project scope went from specification all the way through SoC silicon on a finalized reference board following the steps as outlined below:

- Architecture
- IP qualification
- RTL design, integration, and verification
- FPGA prototyping
- Software development
- Physical design
- Manufacturing, packaging, and test
- HW boards and silicon validation
- Volume production
- IoT gateway reference platform

Voledia Architecture & IP Qualification

Voledia is architected to be an IoT gateway platform that can be targeted to serve the following IoT applications:

- Residential media gateway
- Enterprise wireless AP
- Enterprise security gateway
- Voice gateway
- Smart city gateway

The Open-Silicon design team used their IoT reference platform to do its initial architectural trade-offs. As already mentioned, the Open-Silicon IoT gateway platform is based on the ARM Cortex processor architectures. For this customer's applications, the design team chose to use an ARM Cortex A9 Triple Core running at 1GHz +, along with a 32KB L1 cache, a 256KB L2 cache, a Snoop Control Unit, and the ARM NEON and VFP subsystems.

The platform also used several interfaces and IP blocks, including the Arteris FlexNoc interconnect fabric, AXI, a power management unit, and drivers for DDR3, PCIe 2.0, USB 2.0/3.0, SATA 3.0, GMAC, MMC, NOR/NAND, SPI, I2S, I2C, PCM/TDM, UART, JTAG, GPIO, and timers. The architecture also included customer supplied custom proprietary IP blocks for packet processing, cryptography, PTP, TOE, and the PMU.

SoC architectural analysis and micro-architectural tradeoffs were done using ARM Virtual Prototype tools. Trade-offs were made for both performance and power. Power modes were handled by a hardware-based PMU and finite state machine. The SoC supported four power domains with six clock domains and five system power modes. The PMU and RTC were part of the "Always On" domain. The DDR3 interface supported a self-refresh capability while in standby modes.

Performance tuning was enabled using CPU overdrive by pushing voltages up from 1.1V to 1.2V. I2C controls were available for turning voltages on or off, and for varying voltage levels. The SoC also had an on-chip thermal diode used for temperature sensing. PWM was used for fan control. Digital ring oscillators were also used for process variation sensing.

Custom SoCs for IoT: Simplified

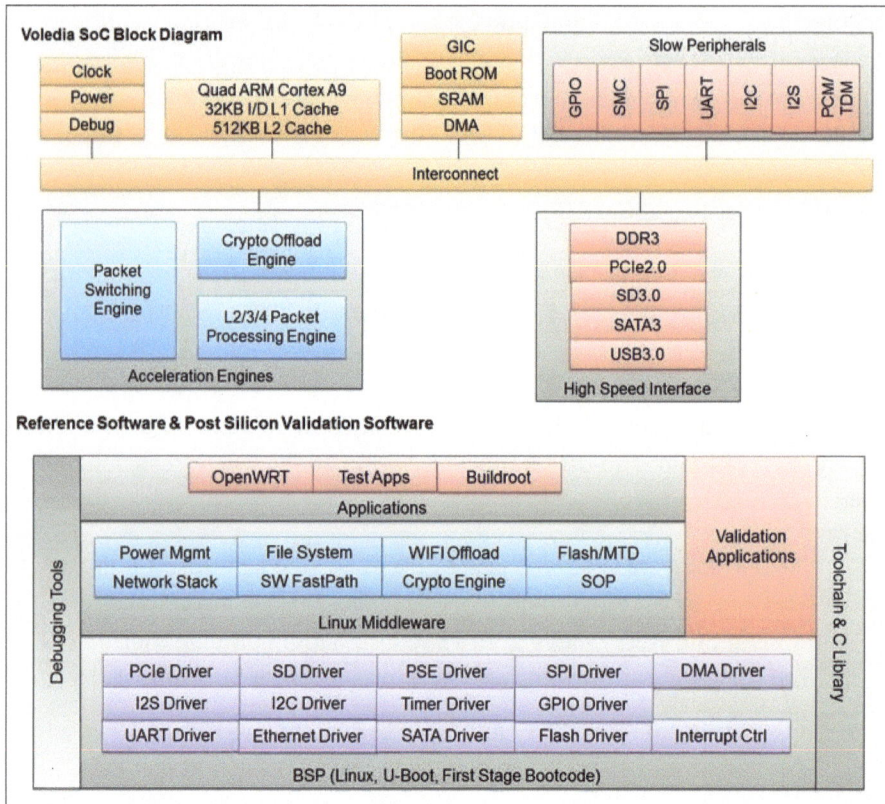

SoC block diagram including micro-architecture and software stack. Image courtesy of Open-Silicon.

RTL Design, Integration & Verification

Much of the RTL code was synthesized using the ARM Virtual Prototype tools. However, custom RTL code was also written for clocking, reset, and PMU logic. The RTL was verified using high-level models to create a virtual prototype of the system and then re-verifying the virtual prototype against the RTL. Verification IP for the ARM processors, device drivers, and Arteris interconnect fabric were used during the verification process to ensure the design was properly configured as compared to the ARM Virtual Prototype behavioral models.

FPGA Prototyping & Software Development

The IoT SoC was prototyped using a custom engineered FPGA prototype board designed by Open-Silicon that used an A9 CoreTile Express processor and two Xilinx Virtex 6 FPGAs. Overall ASIC complexity included 40 million ASIC gates with an ARM A9 core to implement the SATA 3, PCIe Gen2, DDR3, USB 3.0, SD 3.0, and 1G Ethernet interfaces. This logic was synthesized and partitioned across the two Xilinx FPGAs. A complete Linux software stack, including middleware and applications layers, was developed and validated using this FPGA prototype board.

As part of the overall deliverables, Open-Silicon developed a reference software development kit (SDK) targeted to the IoT residential media gateway ASIC using the FPGA prototyping board. As part of this, they had a dedicated software quality assurance (SQA) team that supported the SDK development and debug on the final SoC silicon.

Physical Design, Manufacturing, Packaging and Test

Open-Silicon synthesized, placed, and routed the SoC using their standard tool flow using design constraints as generated from the architectural phase. The design was taped out and processed using a TSMC 40nm LP (low power) process. As part of the implementation phase, Open-Silicon also developed manufacturing test vectors which were used during manufacturing test.

HW BOARDS & SILICON VALIDATION

While the SoC was being manufactured, Open-Silicon prepared multiple boards, including a silicon validation board and reference boards, to be used by the customer for their end markets.

Upon receipt of manufactured and packaged designs, Open-Silicon ran through their standard silicon bring-up tests using their silicon validation board. Validation steps included functional validation of the supported interfaces, including PCIe, USB, Flash, I2C, SPI, I2S, GPIO, UART, SMC, SDIO, and DDR3. Validation was also performed on the customer's IPs for the packet processing and switching engine, the cryptography engine, and PMU components.

System-level validation was also performed across the different power management modes (e.g., different power/clock domains with dynamic frequency and voltage scaling). The entire system was validated using u-boot and Linux. Validation runs were made for different boot mode configurations. Finally, Open-Silicon also did performance analysis for the IoT gateway SoC along with regression and stress testing.

As mentioned earlier, two reference boards were built for their customer. One was a 250mm x 135mm IoT residential media gateway reference board with 1100+ components. The other was an IoT smart city gateway reference board that can be used by end-market customers to evaluate the Voledia SoC.

Custom SoCs for IoT: Simplified

IoT residential media gateway reference board. Image courtesy of Open-Silicon.

Custom SoCs for IoT: Simplified

SMART CITY IOT GATEWAY PLATFORM BASED ON VOLEDIA SOC

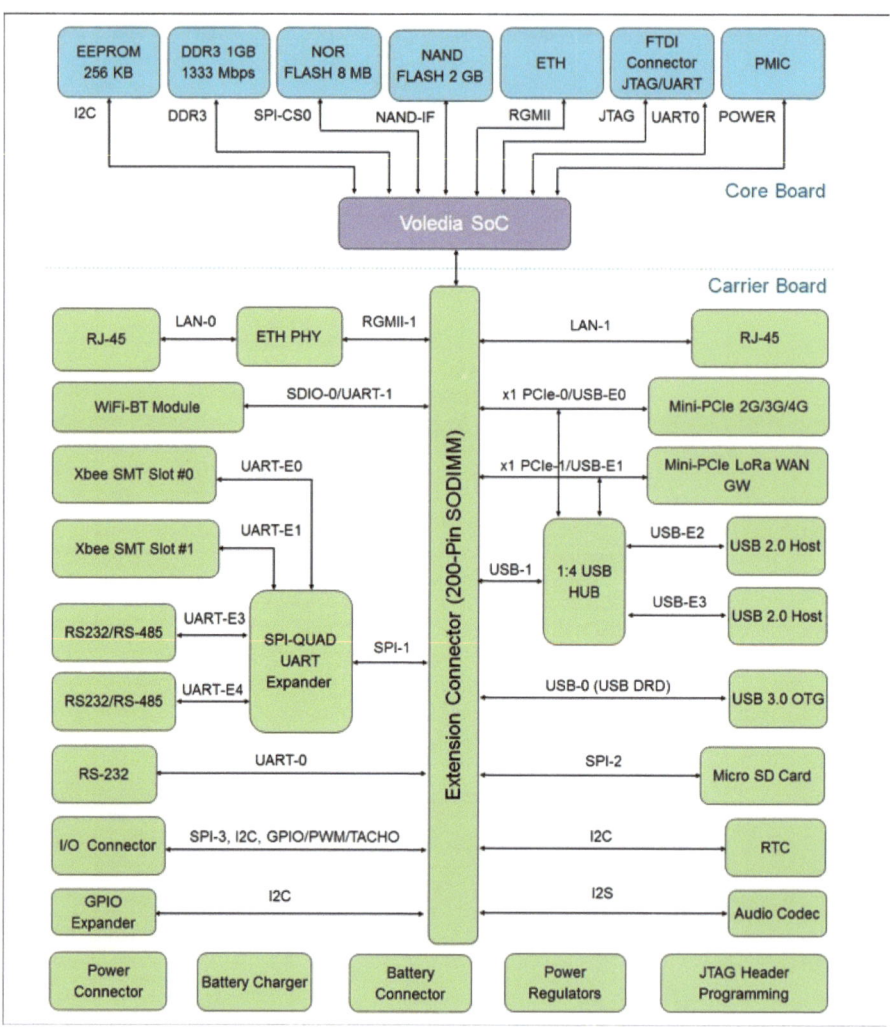

Smart city gateway hardware specification. Image courtesy of Open-Silicon.

Custom SoCs for IoT: Simplified

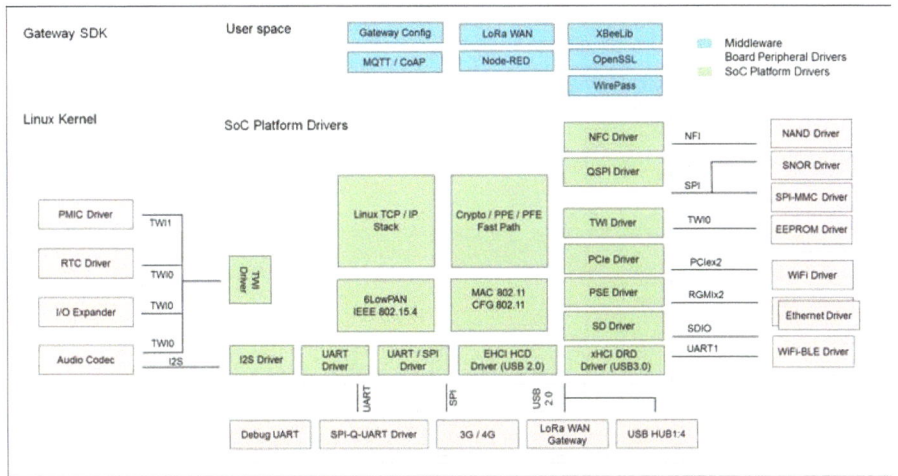

Smart city gateway software development kit. Image courtesy of Open-Silicon.

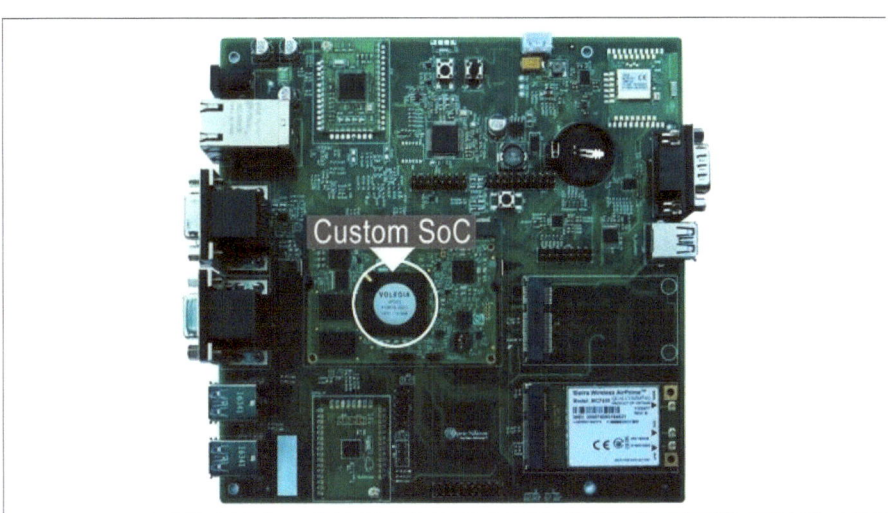

Smart city gateway platform reference board. Image courtesy of Open-Silicon.

EPILOGUE

We are fortunate to be living in one of the most amazing and exciting times in the history of our planet. The developments seen in my lifetime alone have been astounding. It was only forty years ago that I remember the rotary dial telephone in my parents' home that used what was known as a party line where you shared the phone line with your neighbors. You would pick up the phone to find someone else using the line and you had to wait till their conversation ended before you could make your call.

I remember the first Motorola mobile phones that looked like a fancy and smaller version of the famed walkie-talkies used in the military. We were simply amazed that you could walk around with this thing and make calls. I also remember flying from San Jose, California back to Dallas, Texas and having a conversation with a gentleman from a young startup called AMD, telling me about this new thing called the world wide web. Cellular data access and smart phones came along, and the internet seemed to literally explode into existence overnight. Email became ubiquitous and then even more so the advent of Google, where you could find anything on the internet, and the beginnings of social media. I remember as a father with teenagers, my trepidation about allowing my daughters to have Facebook accounts, worried that perhaps the world was intruding into our family life at perhaps too fast a pace. I dug in my heels only to have my wife show me her Facebook page and how she kept track of what our kids were doing by friending them. She now routinely does the same thing with our grandchildren.

As we write this book, we are on the cusp of yet another inflection point. The world wide web has morphed into the internet of things, some even call it the internet of everything, and this book briefly touched on some of the technical aspects of how people are working to bring products to the market to exploit these new opportunities. But, as said, we are just at the beginning. The IoT is emerging just as we see an explosion of progress in the fields of artificial intelligence and virtual reality, which should, in fact, enable many of the autonomous systems-of-systems talked about in this book. This too will be a stepping stone to even more aggressive technologies that are already being developed in research labs.

As an example, silicon photonics is on the cusp of becoming a mainstream technology that will enable much of the 5G cellular network that will truly make the IoT ubiquitous. Interestingly, photonics is also an enabling technology for things like neural networks that operate at the speed of light and quantum computing which will radically change the way in which we do computation. The lessons learned and the platform-based methodologies used by companies like Open-Silicon will be key to enabling companies both large and small to manage future complexities and to move the state-of-the-art forward. At this point, I'd love to make a reference to Dick Tracy and/or the Jetsons but my guess is that the reference would be lost on most readers. If you are curious, look them up and be amazed at how much of the vision of the future has come to pass. Then write your own future and go out and do it. There truly is nothing stopping you.

Mitch Heins
October 2017

ABOUT THE AUTHORS

Daniel Nenni has worked in Silicon Valley for the past 30 years with computer manufacturers, electronic design automation software, and semiconductor intellectual property companies. He is the founder of SemiWiki.com (an open forum for semiconductor professionals) and the co-author of "Mobile Unleashed: The Origin and Evolution of ARM Processors in Our Devices," "Fabless: The Transformation of the Semiconductor Industry," and "Prototypical: The Emergence of FPGA-based Prototyping for SoC Design." Daniel is an internationally-recognized business development professional for companies involved with the fabless semiconductor ecosystem.

Mitch Heins has 34 years of experience in the semiconductor and design automation (EDA) industries. He held management positions in every part of the IC design ecosystem and helped birth the ASIC and EDA markets in the 1980s. Mitch has worked in both startups and large companies and has hands-on experience in logic synthesis, static timing analysis, logic simulation, schematic capture, custom layout, auto-place and route, and design-for-test (DFT). He also pioneered efforts in design-for-manufacturing (DFM), working with leading equipment manufacturers for mask writing and wafer manufacturing. More recently, Mitch has been involved with the newly emerging integrated photonics industry.

www.ingramcontent.com/pod-product-compliance
Lightning Source LLC
Chambersburg PA
CBHW040223220526
45473CB00001B/95